Zinfandel

A Reference Guide to California Zinfandel

BY CATHLEEN FRANCISCO

THE WINE APPRECIATION GUILD
SAN FRANCISCO

Published by
THE WINE APPRECIATION GUILD LTD.
360 Swift Avenue
South San Francisco CA 94080
Phone: 800 231-9463, FAX 650 866-3513
Email: info@wineappreciation.com
Internet: WWW.wineappreciation.com

ISBN 1-891267-15-9

Library of Congress Catalog-in-Publication Number

Maurice T. Sullivan, Editor
Therese Shere, Index
Woodland Graphics, Cover Design
Mimi Osborne, Cover Painting

For Jesse & Elizabeth

son my sun, daughter my moon

Foreword

BY JOEL PETERSON

When I first crushed grapes for Ravenswood in 1976 there were only a few producers of excellent Zinfandel. Wine industry pundits suggested that Ravenswood would be unlikely to survive producing only red wine, and certainly would never survive if that wine was Zinfandel. "America doesn't drink Zinfandel," I was told.

But producing Zinfandel made sense to me, after all it was California's most widely planted grape. I had noted that the best of old-vine Zinfandel was grown like the grapes for great French growths: dry-farmed in exceptional locations, using a special trellising system (head pruning) that allowed for light penetration and air flow, the vines were closely spaced and had low crop production. In addition, I had tasted some memorable older wines made from Zinfandel.

I was intrigued with the grape and worked with Joe Swan, whose wines from 1968-1975 showed me just how spectacular Zinfandel could be. The added incentive was that Zinfandel grapes were inexpensive and I didn't have much money. Never mind that most Zinfandel was not planted in the right place to produce great wine and that the single biggest user of the grape was Gallo before they got into the fine wine business.

If I was successful at my quest, I told myself, I could be a big fish in a small pond. The splash I made with that first production, 327 cases, was not very big or very loud but it was enough of a ripple to allow Ravenswood to grow it's Zinfandel production.

It would not have been hard to list all the Zinfandel producers of note by the late 1980's; there were only a handful. People had begun to focus on the "3R's" of Zinfandel: Ridge, Ravenswood and Rosenblum. There were others: Storybook, Nalle, Martinelli, Gundlach-Bundshu, Dry Creek and the ubiquitous Sutter Home.

A few years later in the 1990's, Zinfandel was still in feeble shape. Some of the larger wineries had dropped it from their production line and most people in the U.S. thought that Zinfandel was a white grape. A group of producers that included Storybook, Ridge, Ravenswood and Rosenblum, got together and formed ZAP: Zinfandel Advocates and Producers. This alliance of producers and consumer advocates began by putting on educational tastings and seminars about America's heritage grape. Much to our surprise we found a dedicated group of closet Zinfandel lovers to join with us in our revelry.

Whether it was ZAP, Cabernet and Chardonnay overload, or just karma, Zinfandel popularity began to explode. By the last ZAP tasting of the millennium there were 125 producers of fine Zinfandel pouring more than 500 wines to the greater than 5000 people in attendance.

Needless to say, the pond that was so small has now become much, much larger with many substantial fish all vying to be the Big Zin of the pond.

I can say that I was more than a little enthusiastic when I heard that Cathleen Francisco was working on a book that would give structure to the now complicated world of Zinfandel. I began reading her book so I would know what the rest of the fish were up to and came away with a renewed enthusiasm and appreciation for this unique grape. The author not only lists every important Zinfandel producer, but gives the reader a sense of place, style and proportion for each winery's Zinfandel.

Just as it has been important to the consumer of fine Burgundy wines to know the locations of, for example, Beaune and Côte de Nuits, the many vineyards of each, and the even greater numbers of producers of each wine, to enhance their pleasure and enjoyment, it has become true for "Zinners" as well. Cathleen Francisco's lovingly assembled book is a fine way to learn and investigate the increasingly complex and increasingly pleasurable world of Zinfandel.

Zinfandel

A Reference Guide to California Zinfandel

CONTENTS

Glossary 319

Terms used in making
and tasting Zinfandel

Blending Varieties 344

Other grape varieties used
as a blending component
in creating Zinfandel

Viticultural Areas 350

Viticultural areas
suitable for creating
Zinfandel in California

Index 359

INTRODUCTION TO
Zinfandel

Zinfandel

The first records of Zinfandel's arrival into the United States from Europe is in the early 1820's where New England nurserymen cultivated the variety, under hot house conditions, as a table grape. Some thirty years later it was brought to the west coast and planted in the vineyards of California. But why was the original vine, a species of Vitis vinifera, labeled Zinfandel when no such variety existed anywhere else in the world? The origins of Zinfandel seem tied to Eastern Europe with DNA studies showing it to be identical to the Primitivo grape of Italy and possibly related to Plavac Mali of Croatia. Yet questions remain about its heritage and those who seem most knowledgeable about such matters still disagree on the particulars of Zinfandel's beginnings. The true source of Zinfandel may always remain a puzzle but there is no question that this orphan red grape, often simply referred to as Zin, now belongs to California.

Prior to Zinfandel's introduction in the early 1850's to California, the Mission grape was the most widely planted variety. A vigorous vine that was relatively resistant to disease, producing large crops and inexpensive to grow, the Mission was selected by monks to make wine for trade and sacramental purposes. But winemakers despaired over the bulk wine the Mission grape produced for it often lacked structure and additives were sometimes required to give it color. The interest in Zinfandel grew when winemakers began blending the variety with the Mission grape, discovering that the wine gained in character and hue. As a varietal Zinfandel grew in popularity producing wines that ranged in style from rosé to robust. The unique characteristics and blending versatility of the Zinfandel grape soon garnered respect, and in 1865, winemakers and growers selected it as one of the five best varieties for the future of winemaking. Demand increased and by the 1880's Zinfandel was a staple in the vineyards of California with plantings exceeding that of the Mission grape.

More than a century has passed and despite its assets Zinfandel has struggled with its standing in California. Wine consumers found Zinfandel's broad range of styles confusing, and its lack of noble heritage contributed to its reputation as nothing more than a good jug wine. The American image of wine was defined by Cabernet Sauvignon and Chardonnay. Sadly, Zinfandel was often removed from the vineyards to make way for those so-called "prestige" vines. The first rosé style of Zinfandel was produced in the 1860's and ironically it was another rosé -blush style wine, known as White Zinfandel, that saved many of the vines from disappearing from the vineyards of California, thus giving Zinfandel a much needed reprieve.

For lovers of Zinfandel the lack of lineage has not influenced or lessened their enthusiasm for the grape. There is a loyalty among Zinfandel fans, an allegiance and a grand passion for the wine. Its versatility is once again gaining respect, not only from winemakers but from wine consumers who find exploring wine a delicious journey.

Nowhere in the world had this grape been grown under the name Zinfandel. Now its name is known, and there is no region in the world where is it as widely cultivated as it is in the valleys and hillsides of California. The orphan grape has found its home and with the help of its patron saint, the organization called ZAP (Zinfandel Advocates and Producers) it may one day be regarded as America's heritage grape.

Zinfandel is considered by many to be one of the more challenging grapes to grow. It breaks bud and flowers early, making it susceptible to frost. Ripening in mid-season, the ample grape clusters are dense with large, thin-skinned berries. Mildew and bunch rot can pose a problem due to the tightness of the clusters, and berries are notorious for ripening unevenly. On one cluster, grapes can range from a pale pink hue to a deep black color, with some berries beginning to over-ripen, turning to raisins.

Zinfandel requires a long growing season with a preference for a warm but not an excessively hot growing region. Hang time is important to insure the delicate balance between acid, tannin, and sugar with the Brix carefully monitored for optimum conditions to begin the harvest. But many winegrowers still rely on experience, tasting the grapes and evaluating color to determine when the berries have reached maturity.

Zinfandel responds well to most soils but tends to produce the best wine quality grapes on well drained slopes. Fertile valley soils and irrigation can cause over-cropping, lessening the flavors of the fruit. Selection of proper root stock, vine training and site protects Zinfandel from climate adversity and if protected, the vines can have an incredible life span, continuing to produce high quality fruit. In places like Amador and Southern Sonoma counties Zinfandel vines that survived the pillage of the 60's have reached a ripe old age of 125 years and more. Many grape growers and winemakers believe this age adds to the grape's intensity. Fruit production lessens but what comes forth is a smaller more concentrated berry, full of focused aromas and broader, richer flavored fruit. This reduced production is not always economical and many of the old, ancient vines are removed from the vineyards. It is a loss, not just of unique fruit quality, but of ancestry and legacy.

Traveling through the vineyards one sees small, stocky vines with wild looking appendages. They look old and forlorn. Craggy and stubborn in stature this style of head training adapts well to Zinfandel. Zin can be a prodigious producer, and some grape growers prefer this type of training and spur pruning to reduce the tendency to overcrop. Spur pruning can also facilitate canopy management, curtailing the risk of sunburn and raisining, producing a more even ripening grape. There are different methods of training a vine, and head training is considered by some to be an old-world approach, a method tried and true. Others disagree, successfully trellising Zinfandel for quality wine production. Terraced vineyards are often trained and pruned to a trellis system. Some grape growers believe this style of training aids in sunlight exposure and air circulation, helping to curtail vine disease, mildew or rot, and is physically easier to harvest.

The characteristics of the wine will vary from region to region but there is a constant with most Zinfandels: a rich, up-front core of fruit, exotic spice and an unmistakable white or black pepper note, that is not just a taste, but a sensation, a texture, like a tickle on the tongue. Sensitive to the soil it is grown in, appellations have a great influence on Zinfandel. It has a remarkable ability to mirror the site it is grown in, as if to take on the personality of a vineyard. It is terroir, the essence of place, that speaks out in the wines made from Zinfandel. Certain regions and vineyard sites produce fruit with varying nuances from chocolate to spice and mint, to a whole spectrum of berries and other red fruits. Winemakers use varying techniques to highlight and frame these nuances. Blending other varieties such as Petite Sirah, Carignane or Cabernet Sauvignon can also be a valuable tool in supporting tannins and fruit flavors while adding other exciting layers of complexity.

The process of vinification and maturation are all reflected in the final style of the wine. Each winemaker approaches his/her wine differently. Each has methods and techniques that they believe are the best for their fruit. Most agree though, minimal handling is ideal. Winemakers approach the grape by first recognizing its character and potential. They want to let the vineyard have its voice, with the wine becoming the minstrel of its terroir.

Zinfandel is a great wine tasting experience. Diverse yet familiar, the wines produced from Zinfandel continue to improve. Grape growers and winemakers alike recognize the importance of vineyard site selections and the techniques utilized in producing Zin enhance, rather than cloak the varietals unique characteristics. For those whose passion is Zinfandel the joy of exploring this variety is always a wondrous, mouth pleasing pleasure. To newcomers of Zinfandel, it is a welcome surprise.

THE
Vintage of 1998

In most of the growing regions of Northern California, the vintage of 1998 was the year when winemakers and grape growers found religion—pleading to omnipotent beings on high was a common and daily ritual. But there is no reasoning with mother nature and the growing season was one of constant trial. This was a season not for the faint of heart.

With the harvest of '97 complete, the vines answered their call to stillness as winter settled in, and for the farmers and winemakers, the slower paced work of the season began. Winter arrived in normal fashion but then decided it did not want to leave. Rain continued for days on end, the ground was saturated beyond capacity and low-lying vineyards were transformed into lakes. And still the rains came. The calendar said it was spring, but winter did not yield with its typical grace and cool days and nights remained in the forecast. Bud break was delayed in some places up to four weeks, and the young shoots that had already pushed were in jeopardy from potential frost conditions. Bloom began under slate colored skies, with cool conditions and rain disrupting the pollination of the flowers, reducing berry set (shatter) in many of the vineyards. With all the stages of vine growth slowed, a late harvest with lower yields was inevitable.

Botrytis typically affects the grapes in fall when damp conditions are more prevalent. But farmers were finding this mold on the young shoots and newly forming berries. More crop was lost and canopy development stunted. Finally summer took over although temperatures for the most part were mild. July was the exception. Temperatures soared into the 100's scorching young berries and leaves, causing the vines to shut down—an act of self preservation. And, once again, fruit was lost. But August was harmless, almost benign as September quickly approached. Now the biggest concern was whether or not there be enough time for the fruit to ripen with the season closing and the vines sensing the cyclical change .

Farmers did not stand idly by and let 'mother' have her way. They spent the season, as they always do, nurturing the vines. Leaf thinning, dropping fruit clusters, holding back water—they used everything they could to help the grapes along. And they succeeded. Although the crop was small, ripened late, and seasonal rains threatened harvest, farmers and winemakers who trusted their instincts and waited were rewarded with ripe, flavorful fruit.

Comparisons are often made from vintage to vintage, but they can be vague, addressing all growing regions, microclimates, and grape varieties with one broad generalization. Nor do these commentaries consider the sagacious farmer or the skill of a winemaker. Each vintage has something unique to offer, and part of the joy of wine is its ability to capture that moment in time. It is for us, as wine lovers, an opportunity to experience those moments, to recognize the challenges that farmers and winemakers have, and savor their passion.

Wine is life, enjoy it!

THE
Wineries

A. Rafanelli Winery

A. Rafanelli Winery
4685 W. Dry Creek Road
Healdsburg, CA 95448
Phone: 707-433-1385

Since the 1950's the Rafanelli family has been growing grapes and making wine in the heart of Dry Creek Valley. Four generations later they are still keeping with tradition, growing premium grapes and producing world class wines. The winery is small and family operated, production is limited, and the focus is on quality, not quantity. The family prides itself on producing consistently exceptional wines year in and year out. The small rustic winery sits on a knoll overlooking the valley and surrounded by some of its hillside vineyards. There are approximately 85 acres in production planted to various varieties.

Tasting daily by appointment between 10:00 and 4:00

Proprietor: Dave and Patty Rafanelli   **Winemaker**: Rashell Rafanelli

The Vineyards:

The Zinfandel grapes come from the benchland and various hillside vineyards located in the Dry Creek Valley. They come from old head trained vines planted on nicely impoverished soils with low yields and lots of hands on coddling. Vineyards are checked frequently for appropriate spacing of the bunches, wings are cut off, as is any second crop, and leaves are pulled to ensure good, even ripening of the berries. The Rafanellis know that in order to have good wine you must first have good grapes! Careful management of the vineyards bring to the winery fruit with depth, ripeness and concentration.

1998 A. Rafanelli Zinfandel

Vinification: The grapes are all hand harvested, then brought to the winery where they are hand sorted, crushed and fermented. Fermentation takes place in open top fermenters where the must is punched down manually to obtain the dark rich color and best tannin balance, then pressed off close to dryness.

Maturation: The wine is aged for 18 months in new to three year old French oak barrels, where it is racked five times during the first year. No fining or filtration of the wine occurs.
Related information:
Alcohol: 14.0%
Residual sugar: Dry
Harvest Date: September 20th through October 5th.
Bottling Date: 3/00
Release Date: 5/00
Production: 6500 cases
Retail: $23.00

Winemaker's Notes:

We focus on high quality and consistency when making our wines. We like to see full-bodied wines that drink pleasant at a young age but benefit from cellar time as well. Our goal is to make a Zinfandel that has dark color, lots of depth, jammy ripe fruit, soft tannins and good balance. The 1998 season was late ripening, due to a cool fall and large crop, producing a soft and supple wine with low acidity. The result is a very smooth and drinkable wine.

The 1998 Zinfandel is full-bodied, rich in color and full of jammy raspberry and blackberry fruit. The spicy and toasty oak flavors linger in the background with the soft tannins. The wines gives a pleasant roundness on the palate, showing elegance and balance from start to finish.

Cellar Notes:

All the wines age wonderfully.

A. Rafanelli Winery also produces Cabernet Sauvignon, Merlot, and Cabernet Sauvignon "Terrace Select".

Acorn Winery

Acorn Winery
P.O. Box 2061
Healdsburg, CA 95448-2061
Phone: 707-433-6440 Fax: 707-433-7641
Email: acorn-winery@excite.com
Tasting and Sales by appointment only.

Acorn Winery's Alegria Vineyards Heritage Vines Zinfandel is a benchmark field blend. Located in our Alegria Vineyards, near Limerick Lane, in the Russian River Valley Appellation, Acorn Winery continues a hundred-year tradition of wines made in the vineyard. The vineyard is mostly Zinfandel with the majority of the vines planted in the 1890's. The balance of vineyard is interplanted with other complimentary varieties which are harvested and fermented together. The ancient low yielding vines produce concentrated flavors of blackberry, raspberry, and spice. This is our second vintage.

Owners: Bill and Betsy Nachbaur: **Winemaker**: Bill Nachbaur

Winemaking Philosophy:

We respect the wisdom of the growers who preceded us and established the vineyard blend, which has evolved over more than 100 years. Being both the grape grower and the winemaker concentrates our attention on the grapes. We strive to produce drinkable, food friendly wines that reflect the vineyard's unique blend and terroir.

1998 Alegria Vineyards Heritage Vines Zinfandel

Appellation: Russian River Valley

Composition: 78% Zinfandel, 10% Alicante Bouschet, 10% Petite Sirah. The remaining 2% of the field blend includes Carignane, Trousseau, Sangiovese, Petite Bouschet, Negrette, Syrah, Muscat Noir, Cinsault and Grenache.

Vinification: The long cool season and small berry size produced intense color and concentrated fruit. We use small, open top fermenters, Brunello yeast, and frequent, gentle punch downs to capture the essences of the fruit. Our mixture of old and new oak barrels ensures that the fruit flavors predominate. We benefited from the expertise and advice of the folks at Rosenblum Cellars, where the wine is made, aged, and bottled.

Maturation: The wine was in barrel for 18 months before bottling in late April 2000, establishing fully integrated and balanced fruit and oak flavors. French, American and Hungarian oak. (22%new, 44% 1 year old and 33% older)

Related information:

Alcohol: 13.9%	**Residual sugar**: Dry
Brix at Harvest:25	**Harvest Date**: October 12 & 14, 1998
Bottling Date: April, 2000	**Release Date**: June, 2000
Production: 280 cases	
Retail: $18.75	

W i n e m a k e r ' s   N o t e s :

The challenging growing season (rain in May, long and cool) produced a smaller than usual crop with highly concentrated fruit. Extended barrel aging resulted in a smooth, complex, and delicious wine.

Acorn Winery also produces Dolcetto, Sangiovese, Syrah, and Cabernet Franc.

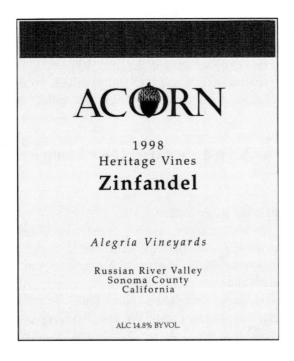

ACⓄRN

1998
Heritage Vines
Zinfandel

Alegría Vineyards

Russian River Valley
Sonoma County
California

ALC 14.8% BY VOL.

Alderbrook Vineyards And Winery

Alderbrook Vineyards and Winery
2306 Magnolia Drive
Healdsburg, CA 95448
Phone: 707-433-9154 Tasting Room: 707-433-5987 Fax: 707-433-0165
Email: info@alderbrook.com
Website: www.alderbrook.com

Alderbrook Vineyards and Winery is located at the southern tip of Dry Creek Valley, near the town of Healdsburg, Sonoma County.
Tasting Room, mail order, website with online order form.
Proprietors: Bruce and David Myers, Tom and Perry Charter, Court King, Clay Shannon and Alen Darr
Winemaker: Kristi Koford

Winemaking Philosophy: Good fruit, good equipment and solid winemaking skills - these are the ingredients that fuel the success of Alderbrook wines. These ingredients are combined with the care and passion of all who grow the grapes and make the wine, creating a magic that enhances mood and food.

1998 OVOC Zinfandel

Appellation: Sonoma County
Composition: 91% Zinfandel, 9% Proprietors Blend
Vinification: Multiple vineyards sources of small berried, loosely hung clusters were harvested, crushed, destemmed, cold soaked, then allowed to warm to 85°F for primary fermentation. The grapes were pressed at dryness, and the wine was barreled while still warm. Malolactic fermentation took place in barrels.
Maturation: The wine was racked twice during the 10 months of barrel aging in American oak, 28% new.

Related information:
Alcohol: 14.2%   **Residual sugar**: Less than 0.1%
Brix at Harvest: 25
Harvest Date: Mid to late October
Bottling Date: 8/99   **Release Date**: 6/00
Production: 10,000 cases sold at winery and distributed.

Alderbrook Vineyards and Winery. Dry Creek Valley   15

Retail: $19.50

W i n e m a k e r ' s   N o t e s : Blackberry and raspberry fruit aromas, spiced black pepper, vanilla, cocoa and nutmeg. Flavors are deep, complex and lingering. Texture is full bodied and firm.

F o o d   p a i r i n g : This full bodied Zinfandel pairs well with spicy Italian foods, grilled and roasted meats, hearty stews, tomato based dishes, and robust cheeses.

Alderbrook Vineyards also produces Dry Creek Valley Chardonnay, Russian River Valley Gewürztraminer, Dry Creek Valley Sauvignon Blanc, Sonoma Valley Muscat de Frontignan, Sonoma County Cabernet Sauvignon, Russian River Valley Mourvèdre, Russian River Valley Viognier, Russian River Valley Pinot Noir, Russian River Valley Syrah, Sonoma County Merlot, and a Zinfandel Port.

Amphora Winery

Amphora Winery
5301 Wine Creek Road
Healdsburg, CA 95448
Phone: 707-431-7767 Fax: 707-431-0258
e-mail: rick@amphora wines.com web: http://www.amphorawines.com

Amphora Winery is a small, hobbit-like place, nestled beneath the barn of a farmer; a grape grower. The rooms of Amphora, reminiscent of European wineries, are dug into the earth and remain constant and cool...the perfect place for wine to live, (and a hobbit or two). Outside the door of this sweet, earthen place, the valley of Dry Creek opens up. Benchlands, mountains, slow slopes of vines, scrub oak, towering spruce and wild mustard can be seen. It is charm, love of life, land and wine incandescent...all held in Amphora. Founded in 1997, Amphora Winery specializes in hand-crafted red wines from Dry Creek Valley.
No tasting room. Tours and tasting by appointment. Mailing list.
Proprietor and Winemaker: Rick Hutchinson

Winemaking Philosophy:
I strive to create wines that are balanced, expressing the vineyard and the true character of the Zinfandel grape: raspberries, strawberries and spice. Focusing only on Dry Creek Valley, where I have been making wines for 15 years, I have a great affinity for this appellation.

The Vineyard:
This vineyard was selected because of my history working with these grapes. I have made wine from here for many years and know and trust the grower, Rich Mount. Located on the west side of Dry Creek Valley, the vines are 35-40 year old, head trained, and grow on a southern exposed hillside. They receive intense heat during the day but are shaded in the evening, giving them a chance to respirate, the combination of which leads to flavor intensity.

1998 Mounts Vineyard Zinfandel

Appellation: Dry Creek Valley
Composition: Zinfandel and Petite Sirah
Vinification and Maturation: The grapes are hand picked and hand sorted using my well-known philosophy: "If you don't want to eat it,

then I don't want to make wine out of it." Up to 25% Petite Sirah is used as a blending component adding structure and backbone. The grapes are cold soaked for 2-3 days to extract smaller tannins creating softer and suppler wines. The juice is fermented in open top stainless steel fermenters and the cap is pushed down daily. The wine is aged in a combination of 33% new French oak barrels and 67% one year old American oak barrels. The wine is racked up to four times during its barrel aging process and is bottled after one year. Egg white fining is used when necessary and the wine is unfiltered.

Related information:

Alcohol: 14.4   **Residual sugar**: Dry
Brix at Harvest: 24   **Harvest Date**: 09/12/98
Bottling Date: 9/12/00   **Release Date**: 3/01/00
Production: 77 cases
Retail: 24.00

Winemaker's Notes:
Deep red in color, the nose offers aromas of plums, black cherry and cola. This wine displays the true characteristics of Dry Creek Valley Zinfandels: spicy, intense flavors of white pepper, cinnamon, strawberry and red currant judiciously blended with well-balanced oak. The middle palate offers incredible strawberry jam flavors combined with vanillin and white chocolate.

Amphora Winery also produces Syrah and Petite Sirah

Baron Herzog Wine Cellars

Baron Herzog Wine Cellars
5965 Almaden Expwy Suite M
San Jose, CA 95120-1771
Phone: 408-997-8350 Fax: 408-997-8342

Royal Wine Company—the parent company of Baron Herzog—a unique, global wine enterprise head-quartered in Brooklyn with wine (and food) interests in California, Israel, France, Italy and Chile. The company roots go back seven generations to when the Herzog Family began making wines 150 years ago.

Royal's Baron Herzog wholly-owned nationally distributed California brand kosher wines is gaining popularity among the general wine consuming public. Royal is both a producer of premium wines in the United States and also as an importer of excellent kosher wines, beers and spirits from all over the world.

Proprietor: David Herzog, C.E.O Royal Wine Corporation
and Nathan Herzog, Exec.
Winemaker: Peter M. Stern

Winemaking Philosophy:
We seek a Zinfandel that highlights the unique aroma of the variety with a balanced oak component. We believe good body and structure are at their best when the grapes are harvested at 23 to 25 Brix enabling us to achieve an alcohol level of about 14.% with a round and soft mouth feel.

1998 Baron Herzog Zinfandel

Appellation: Lodi
Composition: 100% Zinfandel.
Vinification and Maturation: Warm fermentation at 80° F. to 85° F., malolactic fermentation did not take place, moved to barrels following pressing and first racking. Aged in 50% new and 50% one year old American barrels for ten months. The wine was not fined and only lightly filtered before bottling.

Related information:

Alcohol: 14%   Residual sugar: Dry
Brix at Harvest: 23.5   Harvest Date: 9/24/98
Bottling Date: 8/23-24/99   Production: 1500 cases distributed.
Retail: $12.99

Winemaker's Notes:

The 1998 Zinfandel was an outstanding vintage from the Lodi appellation. The grapes were harvested from the small head-trained vines of the Watts Vineyard planted in 1935. The very cool year has yielded a wine with deep color, rich texture and lovely berry, stem notes of the varietal. As the wine was aged in small, new, and one year old barrels, it has wonderful oak bouquet and flavor.

Baron Herzog also produces Cabernet Sauvignon, Chenin Blanc, Chardonnay, Sauvignon Blanc and White Zinfandel.

Beaulieu Vineyard

Beaulieu Vineyard
1960 South St. Helena Highway
P.O. Box 219
Rutherford, CA 94573
Phone: 707-967-5200 Fax: 707-963-5920
www.bv-wine.com
Tasting room open to the public, paid tastings.

At the turn of the century a young Frenchman named Georges de Latour arrived in the Napa Valley with a single purpose: to create wines as fine as any in the world. He purchased a four-acre parcel of unimproved property in the town of Rutherford. Moved by the pristine beauty of the scenery before him, he dubbed the land "Beaulieu" or "beautiful place." In 1909 he realized his dream and produced his own wines for the first time.

In addition to producing table wines, Georges de Latour expanded production of sacramental wines, which enabled the winery to prosper during Prohibition. In 1923 Mr. de Latour purchased the stone winery across from his estate that remains Beaulieu Vineyard's home today. He pioneered the aging of Cabernet Sauvignon in small oak barrels. Rather than ship barrels to market, the custom in that era, he became one of the first vintners in California to bottle wines at the winery to ensure their quality. He also hired a French-trained winemaker, Andre Tchelistcheff, whose innovative viticultural and cellar techniques have revolutionized California wine.
Tasting Room open daily.
Director of Winemaking: Joel Aiken

Vintage Conditions:
A top quality, early and large harvest. A lot of rain began this year, but late winter was unusually dry and unseasonably warm, resulting in early March bud break. The warm spring encouraged growth, and we had nearly perfect berry set. Summer was ideal, with moderate heat and better than average crop load.

Harvest started early. It became clear very quickly that crop levels were above the estimates, as tanks filled alarmingly rapidly. Everything began to ripen at once, causing a gridlock of grapes at the winery. Yet the healthy status of the fruit, and the excellent maturity and incredible fruit intensity at the fermenter promised great things for the future.

1997 Napa Valley Zinfandel

Appellation: Napa Valley. 54% St. Helena and 46% Calistoga
Composition: 100% Zinfandel
Vinification and Maturation: Grapes were gently destemmed and the juice sent directly to the fermenters. Fermentation lasted about one week in open top fermenters with regular pumpovers. The average juice temperature was around 80°F. In order to avoid over-extraction, the juice was drained off just prior to dryness, then pressed off for aging in a combination of American oak tanks and small French and American oak barrels of which 10% were new oak.

Related information:

Alcohol: 14.8%	**Residual sugar**: .5%

Harvest Date: September 10-17, 1997
Bottling Date: September, 1999
Production: 13,000
Retail: $16.00

W i n e m a k e r ' s   N o t e s : A big rich and juicy wine, a Zin not for the faint of heart but for those who love FLAVOR! Ripe, blackberry and bing cherry aromas leap out of the glass. Full, fleshy flavors show rich fruit extract, soft tannins, and a wonderful purity of fruit, while the alcohol adds a distinctive velvety texture. Some moderate youthful tannins suggest that a year or two in the bottle will do wonders. Drink with all manner of full-flavored, but not too spicy, dishes.

Beaulieu Vineyards also produces Cabernet Sauvignon, Chardonnay, Zinfandel, Merlot, and a wide range of other varieties.

Belvedere Vineyards & Winery

Belvedere Vineyards & Winery
4035 Westside Road
Healdsburg, CA 95448
Phone: 707-433-8236 Fax: 707-433-4927
Email: www.belvederewinery.com

Belvedere Vineyards & Winery is a friendly, family-owned winery specializing in artisan wines from exceptional northern Sonoma vineyards. For fifteen years Belvedere has sought out parcels of land in the countryside surrounding the town of Healdsburg, looking for special conditions which signal a great vineyard. Three of California's most coveted grape growing microclimates fall within a few miles of town, and Belvedere now farms ranches in all of these appellations: Russian River Valley, Alexander Valley and Dry Creek Valley.

Tasting Room open daily 10 am to 4:30 pm. Mail order. Wine Club.
Proprietor: William and Sally Hambrecht
Winemaker: Kevin Warren

Winemaking Philosophy:
Here at Belvedere I am lucky to have some of the best Zinfandel vineyards in the Dry Creek Valley to choose my grapes from. I believe that distinctive winemaking requires a certain love of detail. Superior grapes must be harvested at their peak and handled carefully, without excessive manipulation between crush and bottling. Winemaking at Belvedere Vineyards & Winery is hand crafted and adaptable, in order to take advantage of our great Zinfandel grapes.

1998 Zinfandel Dry Creek Valley

Appellation: Dry Creek Valley
Vineyards: The grapes are from the Grist Vineyard on Bradford Mountain (75%) and the Sullivan Ranch (25%) in the Dry Creek Valley. The grapes were chosen from vines ranging from 26 to 90 years old concentrating on a balance of depth and complexity. The Grist Vineyard is high above the valley floor with unique red soil that give the wine intensity and richness. The Sullivan Ranch provides complexity and fresh raspberry fruit flavors and aromas.

Composition: 94% Zinfandel, 6% Petite Syrah
Vinification and Maturation: Cold soak, slow controlled fermentation, gentle cap management, Malolactic fermentation, and wine racked out and back into barrels every three months. Fourteen months in a combination of French and American oak barrels. 30% new oak.

Related information:
Alcohol: 14.2%   **Residual sugar**: dry
Bottling Date: March 28, 2000
Release Date: October 2000
Production: 2525 cases sold at the winery and distributed.

Winemaker's Notes:
The 1998 Dry Creek Zinfandel was chosen from vines ranging from twenty-six to ninety years old. Hints of raspberry, cherry, spice and black pepper tempt the palate with their aromas and distinctive flavors. Belvedere's Zinfandel has a lovely, toasty-oak background that lends depth and complexity. This is a big, luscious, jammy wine that has plenty of fruit and backbone, with a long smooth finish.
Food pairing: Grilled meats, BBQ Hamburger, Lamb, Marinated Pork, Tenderloins, Italian food, foods served with peppers and spice.

Cellar Notes:
1993: Drink now. Mellow and smooth, spicy, good berry flavors.
1994: Drink now or hold up to 1 year. Touch of cedar, herbal spice, pepper and berry flavors and aromas.
1995: Drink now or hold up to 4 years. More robust, tighter, more powerful flavors and structure than previous vintages. More blackberry flavor than previous years.
1996: Drink now or hold up to 1 year. Lighter style with jammy raspberry flavors with toasty oak.
1997: Drink now or hold up to 5 years. High intensity, big flavors, smooth, rich deep berry flavors and aromas, spice and vanilla oak. A great full bodied Zinfandel. Yummy.

Belvedere also produces Chardonnay, Cabernet Sauvignon, Merlot, Syrah and Pinot Noir.

Benziger Family Winery

Benziger Family Winery
1883 London Ranch Road
Glen Ellen, CA 95442
Phone: 707-935-3000 Fax: 707-935-3016

The Benziger Family, producers of Benziger Family, Reserve, and Imagery wines, believes that the nature of great wine lies in vineyard character, winemaker artistry and family passion. At Benziger this means farming and vinifying select vineyards to mine the unique personality of each, winemaking that combines intuition and artistry with attention to detail and a pervading family passion. In its quest for uniqueness through diversity, each year the family produces over 300 lots of grapes from over 60 ranches in over a dozen appellations.

Owner: Benziger Family

1997 Sonoma County Zinfandel

Composition: 85% Zinfandel 10% Petite Sirah 5% Syrah

Vintage: 1997 was a warm growing season that started early and, in general, ended with an early harvest. But because the zinfandel vines from which this wine was produced are up to 100 years old, they have deep root systems that protect them against the warm temperatures and early ripening, harvest of most of our Zinfandel began as normal, at the end of August. The 1997 harvest produced grapes of lovely concentration and intensity.

Vinification and Maturation: Ten lots of grapes from seven vineyards were vinified including those from the Carreras Ranch which has vines between 65 and 100 plus years old, the Taylor Ranch, with 25 to 85 year old vines and Casa Santinamaria Vineyards with vines between 75 and 100 years old. These were culled and individually fermented in small open top fermenters for seven to eight days at 75-88° Fahrenheit with daily punching down. The wine was aged in American oak for 14 months prior to blending and bottling in March of 1999.

Related information:
Alcohol: 15.0%   **pH**: 3.48
Total Acidity: .66

Release Date: June, 1999
Production: 4600 cases
Retail: $21.00

Winemaker's Notes:

Raspberry, cherry, allspice and vanillin toast in the nose. Full and rich with bright berry fruit in the mouth, and a soft, round finish. Excellent now or with up to four years more aging. Best with pork roasts or fajitas.

Beringer Vineyards

Beringer Vineyards
2000 Main Street
St. Helena, CA 94574
Phone: 707-963-7115 Fax: 707-963-1735
Website: www.beringer.com

Established in 1876, Beringer Vineyards enjoys the historic privilege of almost 125 years of exploring the unique characteristics of fruit from Napa Valley estate vineyards. For the past two decades, vineyard manager Bob Steinhauer and winemaker Ed Sbragia have worked as a team to ensure that grapes of the highest quality reach the winery and leave as expressions of the winemaker's art over a wide range of varietals. Beringer is the oldest continuously operating winery in the Napa Valley.

Tasting room opened daily 9:30 a.m. to 6:00 p.m. May through October; 9:30 a.m. to 5:00 p.m. November through April. Tours every half-hour on a first-come, first-served basis. Call 707-963-4812 for details

Winemaker: Ed Sbragia

Winemaking Philosophy:

I learned to make Zinfandel from my father who grew grapes in the Dry Creek area and made wine for our home consumption. He taught me the minimal intervention approach that I continue to follow to preserve the great, ripe, peppery berry flavors we are able to achieve in our vineyards for this classic California wine: pick the grapes at the peak of maturity, keep the barrels topped up, and leave the wine alone to develop and evolve on its own.

1997 Zinfandel

Appellation: North Coast

Vintage: Following a very moderate winter, the 1997 growing season began early on the North Coast, with bud break occurring from 10 to 14 days before normal. Each stage of vine growth and fruit maturation followed this early pattern but because the progression from one stage to the next occurred in the normal number of days - or even longer - the grapes had exceptionally long hang time to develop and evolve in flavor.

Because the temperatures remained moderate for the entire season, the grapes were able to achieve full physiological maturity, with intense fruit flavors and ripe tannins.

Composition: After the wine had completed barrel aging, I blended in a small amount - about 13% - of Petite Sirah, a grape that I believe beautifully complements the Zinfandel's blackberry/black cherry flavors.

Vinification and Maturation: After fermenting the wine to dryness, I aged the wine for 14 months in a combination of French and American oak barrels, accounting for the subtle notes of vanillin and nutmeg/cinnamon in the finished wine. I believe that the spicy tones of the American oak barrels I selected for this wine enhance the spicy, peppery character of the Zinfandel grapes.

Related information:
Alcohol: 13.4%
Harvest Date: September 7 through October 27, 1997
Bottling Date: January 2000
Release Date: Fall 2000
Production: Sold at winery and distributed
Retail: $14.00

Winemaker's Notes:

Zinfandel is our family wine, in part because that's what my father made at home from the vineyards I still live in, and in part because it is so high in what my cousins and I call the 'yummy factor.' By that I simply mean that the wine makes your mouth water, getting you ready to enjoy your food to the fullest. In the spectacular 1997 vintage, this wine had this quality in spades: it's a juicy, bright, mouthfilling wine with great balance and a big, voluptuous finish. - Ed Sbragia

Beringer also produces Chenin Blanc, Gewürztraminer, Sauvignon Blanc, Chardonnay, Gamy Beaujolais, Pinot Noir, Merlot, Cabernet Sauvignon, Cabernet Franc and White Zinfandel.

Black Sheep Winery

Black Sheep Winery
P.O. Box 1851
Murphys CA. 95247
Phone: 209-728-2157 Fax: 209-728-2157 Email: info@blacksheepwinery.com

Black Sheep Winery is located in the Sierra Foothill Gold Rush town of Murphys in Calaveras County. We have been crushing since 1984, making limited quantities of wine from selected vineyards located in Calaveras, Amador and El Dorado Counties. We enjoy being part of an expanding industry here in the county, where wine production began in 1851. We feel we are making our own history with the second "modern" renaissance of winemaking.

Tasting Room open Saturday and Sunday 12-5 and weekdays by appointment. Mail order.

Proprietor: David and Janis Olson   **Winemaker**: David Olson

Winemaking Philosophy:

With our small production of 3500 cases, and our traditional approach, we have an intimate interaction with the wine. Grapes are pitch-forked into a small crusher/destemmer. All our red wines are fermented in small, open top fermenters. The must is punched down by hand twice daily for maximum flavor and color extraction. The wine is pressed in small basked presses, and aged in 60 gallon oak barrels. This hands-on approach, combined with selecting the best vineyards possible in the Sierra Foothills, gives our wine its distinctive style.

1997 Amador County Zinfandel

Appellation: Amador County
Vineyards: The grapes come from one of the larger growers in the Sierra Foothills, Clockspring Vineyards in the Shenandoah Valley.
Composition: 100% Zinfandel
Vinification and Maturation: Small open top fermenters, punch down twice daily. Pressed in small basket presses. Small American oak barrels for 11 months.

Related information:
Alcohol: 13.2%   **Residual sugar**: Dry
Release Date: Spring, 2000
Production: 1407 cases sold at winery and limited distribution

Retail: $14.00

Winemaker's Notes:
Here is Black Sheep's signature zin - rich, big on spice and pepper with lots of blackberry in the nose.

1997 Calaveras County Zinfandel Beckman Vineyard

Appellation: Calaveras County
Vineyard Designation: 100% from Beckman Vineyard in Mountain Ranch
Composition: 100% Zinfandel
Vinification and Maturation: Small open top fermenters, punch down twice daily. Pressed in small basket presses. Aged in a combination of French oak puncheons and American oak barrels.

Related information:
Alcohol: 14.6%   **Residual sugar**: Dry
Release Date: Winter 1999
Production: 760 cases sold at winery and limited distribution.
Retail: $14.00

Winemaker's Notes:
Beckman Vineyard is a 5 acre vineyard producing a full, but smooth Zinfandel that will age 4 to 6 years. This jammy, ripe zin shows wonderful black cherry notes.

Black Sheep Winery also produces Sauvignon Blanc, Cabernet Sauvignon and will soon be releasing a Semillon and a Merlot.

Boeger Winery

Boeger Winery
1709 Carson Road
Placerville, CA 95667
Phone: 916-622-8094 Fax: 916-622-8112

Boeger Winery is located in the Sierra Foothills just east of Placerville where scenic beauty and rich California history have intertwined to create a truly memorable setting. It is here that Greg and Susan Boeger purchased a 70 acre ranch in 1972 and pioneered a modern day winery that has since become synonymous with Sierra Foothill winemaking.
Tasting Room open daily from 10 am to 5 pm. Mail order.
Proprietor: Greg and Susan Boeger
Winemaker: Greg Boeger and Justin Boeger

Winemaking Philosophy:
In the more than two decades since Boeger Winery was founded, we have been committed to only one standard: quality. We have built on our education and experience to produce exceptional wines. We start in the vineyard by growing over 85% of our grapes and maintaining well-balanced crop loads through thinning and pruning. We remove leaves by hand from each vine, exposing the fruit to sunshine in order to obtain higher color and flavor intensity. It is also our goal that each wine produced expresses its own, unique varietal characteristics.

The guiding principle of winemaking at Boeger Winery is to produce only wines of the highest quality. To attain that end, we strive to handle the wines only when necessary, utilizing gentle techniques.

1997 Walker Zinfandel

Appellation :El Dorado **Vineyard**: Walker Vineyard.
Composition: 98% Zinfandel and 2% Petite Sirah
Vinification: 7-10 days fermentation on skins, daily cap push-down
Maturation: 18 months in French and American oak barrels

Related information:
Alcohol: 14.%   **Residual sugar**: dry
Brix at Harvest: 25.6   **Harvest Date**: 9/13/97-9/15/97

Bottling Date: 9/22/99   **Release Date**: 10/99
Production: 1,900 cases   **Retail**: $15.00

W i n e m a k e r ' s   N o t e s : Zinfandel grapes from the Walker Ranch have invariably proven to be of exceptional quality, thereby warranting the production of a vineyard designate wine. Boeger Winery grows over 85% of the grapes used in wine production, the most notable exception being the Walker Vineyards Zinfandel, which constitutes roughly 12% of the total tonnage of grapes brought in by the winery. Our long-standing relationship of over twenty years with grower Lloyd Walker helps ensure that only the highest and most consistent quality grapes ever reach the fermenters. The 1997 Walker Zinfandel is bright, with forward fruit and spice aromas.

1997 Estate Zinfandel

Appellation :El Dorado   **Vineyard Designation**: Estate
Composition: 89 % Zinfandel and 11% Petite Sirah
Vinification: 7-10 days fermentation on skins, daily cap push-down
Maturation: Aged 18 months in small American oak barrels.

Related information:
Alcohol: 14.4%   **Residual sugar**: Dry
Brix at Harvest: 25.3   **Harvest Date**: 9/20/97-10/01/97
Bottling Date: 5/18/99   **Release Date**: 6/18/99
Production: 650 cases   **Retail**: $15.00

W i n e m a k e r ' s   N o t e s : The land upon which Boeger Winery and its vineyards now exist was originally homesteaded by the Lombardo family in the mid 1800's. Until prohibition, the family residence doubled as a fully functional winery and now serves as our tasting room. One variety planted by Giovanni Lombardo in the late 1800's was Zinfandel, some vines of which remain today. In the early 1990's we took cuttings from these vines and propagated them into what we call our "Old Clone" vineyard. The berries and clusters are smaller and less tightly compacted than other clones of Zinfandel grown at the estate. These attributes, which contribute to greater intensity of colors, aromas and flavors, help to distinguish the Estate Zinfandel from others. The 1997 Estate Zinfandel is big and chewy, accentuated by a spicy nose and jammy flavors.

Boeger Winery also produces Barbera, Cabernet Sauvignon, Chardonnay, Meritage, Petite Sirah, Merlot, Pinot Noir, Sangiovese, White Riesling and Sauvignon Blanc.

Bohemian Cellars

Bohemain Cellars/BC Wine
P.O. Box 632
Geyserville, CA 95441
Phone: 707-857-4345 Fax: 707-857-4362
Email: bcwines@sonic.net

Bohemia Cellars was created in 1996 by Lesley Strother and Shon to produce wines that they loved to drink. Since they feel that wine is best enjoyed as a compliment to food, they chose varieties that paired well with their style of cooking. Wines like Zinfandel and Sangiovese that can balance with Southeast Asian, Mexican, Indian or Mediterranean cuisines of North Africa or Southern Europe. Our "Bohemian" moniker references our Gypsy ancestors, our commitment to the Fine Arts, and our unconventional approaches to wine making and marketing. Lesley Strother is an artist, graphic designer and manages B.C. Wines. Shon has been making wine commercially since 1988.

B.C. Wines does not have a tasting room at the present time but we are available by appointment only for tasting and sales.

Owners: Lesley Strother and Shon   **Winemaker**: Shon

Winemaking Philosophy:

We are convinced that small lots of grapes can best represent the terroir from which they spring. By keeping our lots small we can best preserve, undiluted, those unique intense characters of the vineyard. We attempt to get most of our color extraction prior to alcoholic fermentation and use specific strains of yeast to enhance the wines mouth feel. We use new European oak in a very limited way, framing and supporting the focus of the fruit. Gentle treatment, minimal racking for clarity and bottling unfined and unfiltered represent our attempts to reflect the vineyard in the bottle. We use a synthetic, neutral closure in order to guarantee that what we have achieved as wine producers will be there for our customer to enjoy untainted.

1998 B.C. Zinfandel

Appellation: Alexander Valley
Vineyard: All of our Zinfandel comes from the Rancho Rio vineyard, east of the Russian River and just north of Asti. These mature vines were planted on their own roots circa 1950 and are cordoned trained, spur

pruned. Our Petite Sirah comes from an antique head pruned, dry farmed vineyard in southern Alexander Valley. Yield reach 1.5 tons per acre.

Composition: 90% Zinfandel 10% Petite Sirah

Vinification and Maturation: Picked for flavor in the cool of the morning, the grapes were cold soaked for 48 hours before initiating fermentation in 1 ton fermenters. Punched down gently by hand three times a day, they were pressed at dryness. Malolactic bacteria were added to the press tank and completed in barrel. 11 months in 60 gallon European oak barrels, French and Hungarian, 12% new. Unfined and unfiltered and blended prior to bottling.

Related information:

Alcohol: 14.7%	**Residual sugar**: Dry
Brix: 24.3	**Harvest Date**: September 7th, 1998
Bottling Date: July 28th, 1999	
Release Date: March 1st, 2000	
Production: 477 cases	
Retail: $18.00 - $20.00	

Wineraisers Notes:

The 1998 B.C. Zinfandel is our third vintage from Rancho Rio Vineyards and I feel that we are learning how to handle the unique fruit of this two acre block of old Zinfandel vines. After the GONZO harvest of '97, the vineyard yielded fewer bunches with smaller grape berries. The 1998 B.C. Zinfandel can best be characterized by its grace and elegance. The wine has an intense, deep scarlet color with purple highlights. It's luscious berry-pepper aromas rise out of the glass. Wild pungent jam-berries and ripe pomegranate are the dominate flavors. The finish is long, an interweaving of dense, mature fruit with refined oak.

Food pairing: The 98 B.C. Zinfandel can accompany a broad range of foods. It's natural pepperiness is a compliment to spicy foods like curries and chili and is especially partial to barbecue or a mixed grill. If there is anything left in the bottle, well break off a piece of chocolate and let it melt in your mouth. Then finish the Zin!

Bohemian Cellars also produces a limited amount of Sangiovese and "Mountain Merlot.

Carlisle Winery & Vineyards

Carlisle Winery & Vineyards
1623 Willowside Road
Santa Rosa, CA 95401-3965
Phone: 707-566-7700   Fax: 707-566-7200
Email: wine@carlislewinery.com

     Carlisle Winery & Vineyards is a dream come true for longtime home winemakers Mike and Kendall Officer. Their goal is to produce world-class, distinctive Zinfandels from some of the finest old-vine vineyards in Sonoma County, including their Russian River home ranch planted in 1927 by Alcide Pelletti.

Owners: Mike Officer and Kendall Carlisle Officer
Winemaker: Mike Officer

Winemaking Philosophy:

Our philosophy is that great wines are made in the vineyard. With that in mind, we have sought grape sources that we believe produce exceptional and unique grapes. Once the grapes arrive to the winery, our goal is to nurture them into wine as gently as possible so as to preserve their uniqueness and special sense of place.

The Vintage:

1998 was one of the most challenging growing seasons in recent decades. Abnormally high rainfall during spring led to heavy mildew pressure in the vineyards. Growers had to be diligent with their fungicide programs in addition to keeping canopies open through leaf removal. The rain eventually stopped but temperatures remained significantly below average throughout summer. Fortunately, three weeks of perfect Indian Summer weather arrived in early October. With yields significantly down due to the rain at bloom, it was just enough heat to fully ripen our grapes.

1998 Dry Creek Valley Zinfandel

Composition: 76% Zinfandel 24% Petite Sirah

From Ray Teldeschi's ranch, 109 year-old vines and low yields combined to produce some very ripe Zinfandel in 1998. Excited by the intense fruit flavors but concerned that the elevated sugar levels could cause a problematic fermentation, we also picked some flavorful, though less ripe Petite Sirah from an adjacent black and did a "field blend" in the fermenter. The decision paid off and after a long, moderate fermentation, the wine finished dry.

The wine was fermented in small, open-top fermenters using a variety of commercial and indigenous yeasts. Punch down of the cap was by hand. After 18 days on the skin, the wine was gently pressed into a mix of French and American oak, 30% of which was new. No pumps were ever used, only gravity or positive pressure. The wine was racked twice the following year before being bottled unfined and unfiltered.

Related information:
Alcohol: 15.8%   **Residual sugar**: Dry
Harvest Date: 10/08/98   **Bottling Date**: 02/12/00
Release Date: 03/20/00
Production: 232 cases sold at winery and distributed on a limited basis
Retail: $28.50

W i n e m a k e r ' s   N o t e s :
This is an exceptionally rich, decadent Zinfandel with dark color, low acidity, and copious black cherry aromas and flavors. It speaks clearly of Dry Creek Valley. It should be at its best 2 to 4 years from the vintage date.

1 9 9 8   R u s s i a n   R i v e r   V a l l e y   Z i n f a n d e l

Composition: 94% Zinfandel 6% Alicante Bouschet
From Ben Montafi's vineyard planted in 1926, the grapes for this wine were picked at 26° Brix with perfect acidity. It was fermented with indigenous yeasts in small open top fermenters and punched down by hand frequently and vigorously. After 19 days on the skins, it was pressed into a mix of new (40%) and used French oak. It was racked three times the following year prior to being bottle by hand, unfined and unfiltered, in January 2000.

Related information:
Alcohol: 15.9%   **Residual sugar**: Dry
Harvest Date: 10/07/98   **Bottling Date**: 01/15/00

Release Date: 03/20/00
Production: 110 cases sold at winery and distributed on a limited basis
Retail: $28.50

W i n e m a k e r ' s   N o t e s :
Although I prefer to consume Zinfandels young, I feel this wine has the structure and balance to develop additional complexity in the bottle. It's a very serious Zinfandel, offering intense dark color, aromas of vanilla and blackberries, an explosive mid-palate, and a long peppery finish. It should be at its best 3 to 5 years from vintage date but will certainly hold at its plateau for several years beyond that.

Carlisle Winery & Vineyards also produces Petite Sirah, Syrah and a proprietary Mourvèdre blend called "Two Acres".

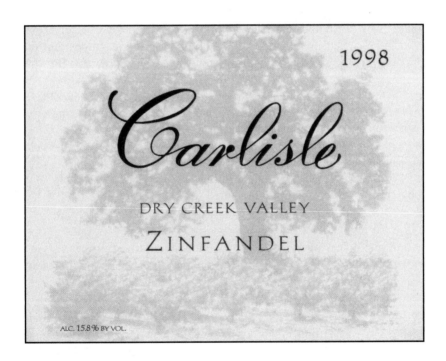

Castle Vineyards & Winery

Castle Vineyards & Winery
1105 Castle Road
Sonoma, CA 95476
By appointment only
Phone: 707-996-1966 Fax: 707-996-1582
Website: www.castlevineyards.com
Email: info@castlevineyards.com

Vic McWilliams is the winemaker and founder of Castle and has been perfecting the art of winemaking and grape growing for over twenty years. After becoming a pharmacy graduate in 1974 from NDSC, Vic decided to make a name for himself in the California wine industry. In 1978, shortly after moving to Sonoma, Vic was invited to a friend's house to help make a barrel of wine. It sparked an interest that developed into a hobby of growing grapes and making wine. Encouraged by awards won in amateur sweepstakes, Vic went "pro" in 1993.

The winery is located on a quiet country road close to Sonoma's downtown Plaza. Vic, Erin, and Buck lead tours through the winery and conduct side down tasting of current release wines by appointment only. Occasionally, an open house will be held on weekends. Call Erin for a schedule.

Founder and Winemaker: Vic McWilliams
Director of Retail Sales, Marketing and Public Relations: Erin McClary
Cellar Dog and Mascot: A yellow lab named 'Buck'

Winemaking Philosophy:
Castle is a minimalist winery. The wines are made in a true European style using old-world techniques like hand punch-downs and open top fermentation combined with modern conveniences.

The Vineyards: Castle specializes in small lot, premium fruit owned or managed by Vic McWilliams. With a small, dedicated crew, Vic manages 75 acres of estate vineyards throughout the Sonoma Valley and Los Carneros appellations. Castle conducts ongoing field trials in the vineyards studying trellis systems, rootstocks and clonal selections.

1997 Zinfandel, Sonoma Valley

Vineyard: From five vineyards, several up to 60 years old, mostly dry farmed.
Composition: 90% Zinfandel with 10% Petite Sirah blended.
Vinification and Maturation: Fermented in open top fermenters. Hand punch-down. Non-fined, non-filtered. Aged in French oak barrels, 35% new.

Related information:
Alcohol: 13.9%
TA: 0.58
pH: 3.6
Release Date: August, 1999
Production: 300 cases
Retail: $19.00

Winemaker's Notes:
Rich black fruit is layered with spice and oak accents. This is a classic, bigger style Zinfandel that will pair well with grilled meats.

1997 Zinfandel, Russian River Valley

Vineyard: Knecht, Vera Gold
Composition: 100% Zinfandel
Vinification and Maturation: Fermented in open top fermenters. Hand punch-down. Aged in American oak barrels, 40% new.

Related information:
Alcohol: 13.8%
TA: 0.60
pH: 3.65
Release Date: January, 2000
Production: 300 cases
Retail: $16.00

Winemaker's Notes:
Ripe red cherry aromas and flavors framed with sweet oak, juicy acids and a soft tannin finish. Pairs well with classic Italian cuisine.

Chateau Souverain

Chateau Souverain
400 Souverain Road
Geyserville, CA 95441
Phone: 707-433-8281 Fax: 707-433-5174

Built in 1973, the winery sits on a hillside just off of Highway 101 at Independence Lane, north of the town of Healdsburg. Chateau Souverain is part of the Beringer Wine Estates group of wineries.

Tasting Room open daily from 10 am to 5 pm. Mail order. Restaurant open daily for lunch, dinner on weekends only.

Proprietor: Beringer Wine Estates   **Winemaker**: Ed Killian

Winemaking Philosophy:

Chateau Souverain makes classic, dry varietal table wines, using only the finest Sonoma County fruit sources. We vinify in a manner that respects the integrity of each variety to produce wines that are representative of their growing region and are unique in their style.

1998 Zinfandel Dry Creek Valley

Appellation: Dry Creek Valley

Vineyards: The Zinfandel grapes used for this wine came from old vines in Dry Creek Valley. The two major vineyard sources for this blend are Buchignani & Garcia Vineyards and the Faloni Vineyard. Most of the vines are head-pruned, and all are planted in rocky, volcanic soils. The Syrah, from our estate vineyard, and Petite Sirah components came from plantings in similar soils, where the vines struggle to grow in the fertile conditions. Low crop yields in these vineyards produce fruit with deep concentration.

Composition: 85% Zinfandel, 10% Syrah, 5% Petite Sirah

Vinification and Maturation: Traditional techniques - including open-top fermentation vessels and seven to ten days of skin contact - were used to extract maximum flavors, aroma and color. After secondary malolactic fermentation, the wine was put up in small French Nevers oak barrels (20% new) for a ten-month aging period. A small amount of

Syrah and Petite Sirah were fermented separately and then blended to enhance the wine's complexity and color.

Related information:

Alcohol: 14.5%	**Residual sugar**: Dry
Brix at Harvest: 24	**Harvest Date**:9/30 - 10/23
pH and TA: 3.61 / 0.60g/100ml	
Release Date: April 2000	
Retail: $13.00	

W i n e m a k e r ' s   N o t e s :
Bursting with jammy boysenberry and black cherry notes, this Zinfandel is enhanced by a rich round mouthfeel, which exemplifies the classic Dry Creek "fruit-driven" style of this wine. Modest acidity and soft tannins deliver a balance, medium bodied wine, touched with back notes of French oak to add complexity and richness throughout the long juicy finish.

Chateau Souverain also produces Sauvignon Blanc, Chardonnay, Cabernet Sauvignon and Merlot, Viognier and a Rhône-style proprietary red.

Christopher Creek Winery

Christopher Creek Winery
641 Limerick Lane
Healdsburg, CA 95448
Phone: 707-431-8243 Fax: 707-431-0183
Website: christophercreek.com

Christopher Creek is a small, family owned winery with a history of producing award winning estate bottled and hand crafted wines. Set in the rolling hills of the northern part of the Russian River appellation, southwest of Healdsburg, Christopher Creek produces Syrah, Petite Sirah, Chardonnay, Zinfandel, Cabernet Sauvignon and Viognier. The vineyard manager, winemaker and owners all work in unison to insure the quality of the grape from vine to bottle. Annual production is 4000 cases.

Tasting room hours: Friday through Monday, 11:00 a.m. to 5:00 p.m.

Propretors: Pam and Fred Wasserman **Winemaker**: Sebastien Pochan
Vineyard Manager: Kerry Damon

Winemaking Philosophy:
Picked ripe but not too late in order to preserve the varietal character. Minimum handling with a focus on fruit and balance.

1998 Zinfandel

Appellation: Dry Creek Valley
Composition: 100% Zinfandel
Vinification and Maturation: 5 days of cold soaking, twice daily punch downs in small open top fermenters with 15 day skin contact. 9 months aging in American oak, 20% new.

Related information:
Alcohol: 14.9%
Residual sugar: Dry
Harvest Date: October 1998
Bottling Date: September, 1999
Release Date: January, 2000
Retail: $22.00

Winemaker's Notes:

The 1998 Dry Creek Zinfandel is the maiden vintage for Christopher Creek. Classical Dry Creek Zin in style, it exhibits a medium-dark ruby hue, with forward strawberry and raspberry fruit, hints of vanilla, spice and mouth feel that lingers forever. While still quite young, this intense, perfectly balanced and bright wine is remarkably approachable with medium tannins. The nose blossoms like spring and is enchanting.

Cline Cellars

Cline Cellars
24737 Arnold Drive
Sonoma, CA 95476
Phone: 707-935-4310 Fax: 707-935-4319

Cline Cellars estate vineyards and tasting room are located at the southern tip of the Carneros Valley. The vineyards produce the finest Rhone-style varietals and Zinfandels in the region. Fred Cline, owner and president, purchased the property in 1989 and the tasting room opened in 1991. Drawing upon extensive vineyard holding in the Contra Costa AVA, winemaker Matt Cline has garnered Cline Cellars enormous acclaim for its drop-jaw Zinfandels.
Tasting Room Mail order. Wine Club.
Proprietor: Fred Cline   **Winemaker**: Matt Cline

The Vintage: 1998 will be remembered as the vintage where "patience is a virtue" was our motto. Compared to the large and early harvest of the previous year, 1998 would be considered a light and late harvest, our first vineyard ripened on September 4th. In fact, the 1998 harvest was two to three weeks later then normal and four to four and a half weeks later then 1997! With an unusually cool spring, the growing season started with what was essentially, very lousy weather. This translated to a poor crop set and a generally low crop size, with the result of lower than average yields. Fortunately, a late summer - early fall heat spell ripened our grapes to maturity.

1998 California Zinfandel

Vineyard: The 1998 edition of the California Zinfandel draws from an even wider assortment of old vine vineyards. A backbone of classic Contra Costa County, fruit grown on low-yielding, head-pruned vines is the unique sands soils of the Oakley area gave this wine an intensity marked by dusty wild berry flavors and peppery tannins. Additional fruit from San Benito, San Luis Obispo and Paso Robles further lend the wine a complex personality and fill out its fleshy, elegant profile. Finally, fruit from the much maligned and under-appreciated Lodi area has brought an explosive berry/cherry character to this Zinfandel.

Vinification and Maturation: Destemmed, lightly crushed with a cool to moderate fermentation in closed stainless steel tanks with selected cultured and wild yeasts. Wines are pressed at dryness and free run and press wine is combined. Aged in French and American oak. Gently filtered.

Related information:
Alcohol: 14.0%   **Retail**: $10.00

W i n e m a k e r ' s   N o t e s : This year's wine has slightly more fruit structure while being intense and nicely balanced with oak. Several selection criteria and master lending show in the final product, which showcases a wide array of dark berry fruit including black cherry and raspberry. Additionally, spice notes and a lasting finish of vanilla from oak aging and firm, supple tannins add complexity to a wine whose average vine age is, remarkably, more that 50 years old. That said, the 1998 California Zinfandel is ready to drink now and will continue to develop complexity over the next three to four years.

1997 Ancient Vines Zinfandel

Vineyard: The 1997 Ancient Vines Zinfandel draws from a wide selection of our oldest, most historic and shyest-bearing Zinfandel blocks in Contra Costa County. Planted by our Italian and Portuguese forefathers in the sandy, phylloxera resistant soils of Oakley, these ancient, dry-farmed vineyards - some over 100 years in age - consistently produce fruit of stunning concentration. The lots that we hand-select for the 1997 blend showed an added degree of ripeness and dimension that is the result of sensitive farming practices, the singular Oakley terroir and a unique cooling band of air that flows from the San Joaquin and Sacramento Rivers.

Vinification and Maturation: The grapes underwent near total destemming and a very gentle crushing to ensure a large proportion of whole berries in the must - contributing to the explosive fruit character of the wine. Fermentation was carried out in small stainless steel with select cultured yeast after a 48 hour cold soak to extract color and flavor early on. The wines were pressed off their skins at dryness and racked gently before being laid down to a compliment of small American and French cooperage - approximately 25% new American. Before bottling, small amounts of Mourvèdre and Carignane were blended with the Zinfandel to achieve a harmonious and structured wine capable of aging for several years.

Related information:
Alcohol: 14%   **Retail**: $20.00

W i n e m a k e r ' s   N o t e s : The 1997 Ancient Vines Zinfandel truly shows the wonderful dusty berry, white pepper and chocolate character that is so particular to Oakley Zinfandels. Prolonged aging in new and used wood has lent this wine a subtle vanilla quality that complements nicely the explosive fruit notes.

1997 Bridgehead Zinfandel

Vineyard: Bridgehead Vineyards, Contra Costa County.
The Bridgehead Vineyard, named for the Bridgehead Road that runs adjacent to this treasured black, consistently produced one of our most individual and refined lots of Zinfandel. Planted by Italian immigrants well before the turn of the century, the Bridgehead Vineyard in Oakley is among this country's most historic. Ancient, over 100 year old head trained vines, dry farming and sandy soils combine with a unique band of cooling air from the Sacramento and San Joaquin Rivers that favors the Bridgehead Vineyard to create an incredible synergy of element for expressing the unique character of this site.
Vinification and Maturation: In 1997, grapes from the Bridgehead Vineyard were harvested at high natural sugar levels and treated in the cellar with kid gloves to emphasize the explosive fruit characteristic of the block. The fruit was almost entirely destemmed and lightly crushed, with a large proportion of whole berries remaining in the must. Fermentation occurred at moderate temperatures in two stainless steel tanks, using a selection of cultured yeast in one and allowing a wild yeast fermentation in the other. After 12 days of gentle pump over and near dryness, we drained and pressed the wine from its skins. Pressed off its skins at dryness, the Bridgehead Zin was minimally handled before being put down to barrel, further accentuating the power and freshness of its dark fruit. The wine was aged for 14 months in a combination of new American (approximately 25%) and used oak before bottling. A small amount of Mourvèdre was added to the final blend for structure and back bone.

Related information:
Alcohol: 14%   **Retail**: $28.00

Winemaker's Notes: We consider the Bridgehead Zinfandel to be our "feminine" Zinfandel. Made in a gentler style, the Bridgehead is consistently a supple and seductive wine that retains the power and richness of this unique vineyard. Balanced with a small amount of Mourvèdre, it is subtle yet bold at the same time. Our 1997 Bridgehead Vineyard Zinfandel is firm and deep with ripe, dusty wild berry character, chocolate aromas and spicy pepperiness. Benefiting from a fine-grained structure of balanced acids and tannins, this wine will age beautifully for well over the next 5 to 7 years.

1997 Big Break Zinfandel

Vineyard : Big Break Vineyard, Contra Costa County

The Big Break Vineyard, named for Big Break Road which runs adjacent to the block, has traditionally produced one of our most powerful and individual lots of Zinfandel. An early ripener, Big Break Zinfandel is year in and out among the first lots of grapes that we pick and 1997 was no exception. The combination of extremely sandy, well drained soils, dry farmed century old head trained vines and the unique band of cooling air from the San Joaquin and Sacramento Rivers that favors Oakley's best vineyard sites, create a synergy of elements that is ideal for ripening Zinfandel and expressing the full character of the fruit.

Vinification and Maturation: In 1997, grapes from the Big Break Vineyard were harvested at natural levels and treated gently in the cellar to emphasize the explosive fruit characteristic of the block. The fruit was almost entirely destemmed and lightly crushed, with a large proportion of whole berries remaining in the must. Fermentation occurred at moderate temperatures in stainless steel tanks, using a selection of cultured yeast. After 14 days of gentle pump over and near dryness, we drained and pressed the wine from its skins. The Big Break Zin was minimally handled before being put down to barrel, further accentuating the power and freshness of its dark fruit. The wines was aged for 16 months in a combination of new and used oak before bottling. A small amount of Mourvèdre was added to the final blend of structure and back bone.

Related information:
Alcohol: 14%   **Retail**: $28.00

Winemaker's Notes: Here, at the winery, we feel that Zinfandel is perhaps the only grape in California that merits being bottled as a vineyard designated lot. Much in the same way as the famed Pinot Noirs of Burgundy, certain old vine Zinfandel vineyards have shown such unique and individual character that they could be considered California's true Grand Crus. The Big Brake vineyard definitely is one such vineyard site. Always one of our most robust wines, the 1997 in particular oozes power, revealing a dark wild berry, exotic spice, chocolate and ripe peach character. A distinct dusty berry character (reminiscent of picking wild black berries in the hot summer sun), what we affectionately describe as our "Oakley terroir," effuses predominately from this wine. The Big Break Zinfandel, which can be enjoyed young, should also benefit from 5 to 7 years of cellaring.

1997 Fulton Road Zinfandel

Vineyard: Our first Russian River Zinfandel is from a 102 year-old vineyard located in the north corner of the Russian River Valley AVA in Sonoma County. Planted on a deep loam soil, age has relegated the yields to a mere ¾ tons per acre. What our Fulton Road Zinfandel lacks in yields is more than made up by the incredible concentration of its fruit. The vineyard is a field blend of approximately 85% Zinfandel, 9% Petite Sirah and 6% Alicante Bouschet. The Petite Sirah provides for the firm tannins that make this wine a monster and a great candidate for cellaring.

Vinification and Maturation: The grapes from Fulton Road were harvested at high natural sugar levels and treated gently in the cellar to emphasize the explosive fruit characteristic of the block. The fruit was almost entirely destemmed and lightly crushed, with a large proportion of whole berries remaining in the must. Fermentation was carried out in stainless steel with select cultured and indigenous yeast. After 14 days of gentle pump over and near dryness, we drained and pressed the wine from its skins. The Fulton Road Zinfandel was minimally handled before being put down to barrel, further accentuating the power and freshness of its dark fruit. The wine aged for 14 months in a combination of new and used oak before bottling - approximately 30% new American.

Related information:
Alcohol: 14%   **Retail**: $30.00

Winemaker's Notes: When a winemaker stumbles upon a vineyard this old and unique which is capable of producing concentrated outstanding fruit such as the Fulton Road Vineyard, it instantaneously becomes a majestic bond. From the moment fermentation was complete, it was obvious that this Zinfandel was destined to become a vineyard designated wine. We are excited to release this first vintage of Fulton Road Zinfandel, with many more to follow. Deep raspberry and blackberry flavors dominate this Zinfandel, with a long, firm peppery finish. Enjoy this Zinfandel now or cellar for 7 to 10 years.

1997 Live Oak Zinfandel

Vineyard: Live Oak Vineyard, Contra Costa County
The Live Oak Vineyard, named for Live Oak Road which runs adjacent to the block, has consistently produced one of our most complex, concentrated and unique lots of Zinfandel. The 100 plus year old vines planted on their original roots make this one of California's most historic vineyards. Cline Cellars has chosen to dry farm the ancient, head pruned vines in the Live Oak block, continuing a practice employed by the Italian immigrants that planted this vineyard well before the turn of the century.
Vinification and Maturation: The fruit was almost entirely destemmed and lightly crushed, with a large proportion of whole berries remaining in the must. Fermentation occurred at moderate temperatures in stainless steel tanks, using a selection of cultured yeasts. After 12 days of gentle pump over and near dryness, we drained and pressed the wine from its skins. The Live Oak Zinfandel was minimally handled before being put down to barrel, further accentuating the power and freshness of its dark fruit. The wine was aged for 16 months in a combination of new and used oak before bottling.

Related information:
Alcohol: 14% **Retail**: $28.00

Winemaker's Notes: This wine has always shown an added dimension of ripeness and individuality, so much so that we decided to bottle it as a vineyard designated wine starting in 1995. A classic example of the big boned Zinfandels for which Cline Cellars has become known, our 1997 Live Oak displays tremendous extraction, a firm and balanced structure of acid and fine grained tannins and a sweet, ripe raspberry flavor.

Cline Cellars produces many Rhone-style varietals as well as three Rhone style blends. Jacuzzi Family Vineyards Zinfandel is also available in limited quantities.

Clos du Bois Wines

Clos du Bois Wines
P.O. Box 940
Geyserville, CA 95441
Phone: 707-857-1651 Fax: 707-857-1667

The Clos du Bois winery is located in the Alexander Valley of Sonoma County. The tasting room is open daily and tours are by appointment only.
Owners: Allied Domecq   **Winemaker**: Margaret Davenport

Winemaking Philosophy:

We select grapes from Sonoma County vineyards near our winery. The winery is opened 24 hours a day during harvest so we can process the grapes as soon as they are harvested. The grape must is fermented in temperature controlled stainless steel tanks and the wine is aged in small oak barrels for 14.5 months. The resulting wine is fruity and easy to drink in its youth and will develop complexity as it ages.

1998 Sonoma County Zinfandel

Appellation: Sonoma County
Vineyard: The grapes are primarily selected from vineyards in Sonoma County's Alexander and Dry creek Valleys.
Composition: 92% Zinfandel, 8% Alicante Bouschet
Vinification and Maturation: The grapes were destemmed but not crushed and transferred to stainless steel fermentation tanks. The alcohol fermentation took place on the skins for 5 to 10 days with a peak temperature of 78F. After alcohol fermentation the wine went through malolactic in barrels and tanks. The wine was aged in 60 gallon barrels, primarily American, for an average of 14.5 months

Related information:
Alcohol: 14.5%
Residual sugar: Dry
Harvest Date: 9/18 - 10/30/98
Bottling Date: April 2000
Release Date: September 2000

Production: 50,000 cases distributed at our winery tasting room and the wine shop nearest you.
Retail: $14.00

Winemaker's Notes:

The 1998 zinfandel crop was well balanced though there was some fruit thinning done in the older head trained vineyards. We had several heat spells where fruit that was overexposed to the sun was sunburned and had to be removed. The crop ripened normally and another heat spell at the end of ripening caused the fruit to come in very ripe and quickly.

The fruit displays a ripe, blackberry plum spice that blends nicely with the smoky vanilla oak from the American oak barrels.

Cellar Notes:

1997 Sonoma County Zinfandel: The fruit displays a ripe, jammy, black pepper spice. There are also hints of raspberry, with vanilla and toasty oak flavors.
1996 Sonoma County Zinfandel: A full bodied wine brimming with ripe raspberry and blackberry flavors with a long peppery finish.
1995 Sonoma County Zinfandel: Medium-bodied with balanced acidity. Soft and lush on the palate
1994 Sonoma County Zinfandel: Toasty, tart cherry, mint and tea flavors with a spicy blackberry finish.

Clos du Bois also produces Chardonnay, Sauvignon Blanc, Merlot, Cabernet Sauvignon, Shiraz and Tempranillo.

Clos du Bois, Alexander Valley

Clos Du Val

Clos Du Val
5330 Silverado Trial
P.O. Box 4350
Napa, CA 94558
Phone: 707-259-2200 Fax: 707-252-6125 Web: www.closduval.com

Founded in 1972 by John Goelet and Bernard Portet, Clos Du Val is located five miles north of Napa on the Silverado Trail, within the Stags Leap District of Napa Valley. With 300 acres of vineyards in four different areas of Napa Valley: Stags Leap District, Yountville, Oakville and Carneros, Clos Du Val is renowned for producing Bordeaux and Burgundian varietals. The name of the winery introduces you to the French heritage of its founders. Clos Du Val translates to "a small vineyard estate in the valley." The label introduces you to the Three Graces: Aglaia (Splendor), Euphrosyne (Mirth) and Thalia (Good Cheer), who embody the winery's philosophy that the celebration of fine wine, fine food and conviviality is truly one of life's greatest pleasures.

Tasting Room and picnic facilities open daily from 10:00 to 5:00. Tours by appointment.
Proprietor: Clos Du Val Wine Company, LTD
Winemaker and President: Bernard M. Portet

Winemaking Philosophy:
Clos Du Val's winemaking philosophy marries a classic, French-inspired winemaking style with the extraordinary terroir of California's renowned Napa Valley, to create wines of the utmost balance, elegance and complexity.

The Vintage:
Thanks to El Niño, 1998 was one of the coolest, wettest years on record, and consequently, Napa Valley experienced one of its latest harvests on record; however, the longer growing season and longer late season "hang time" helped ripening berries reach peak flavors while retaining medium to high acidity. This longer hang time also resulted in wines with above-average richness and color. While the extreme weather conditions, particularly during bloom, caused a lighter crop load, this resulted in some richly concentrated flavors and high-quality fruit.

1998 Zinfandel, California

Clos Du Val, Napa Valley   52

Appellation: California
Composition: 100% Zinfandel
Vinification and Maturation: Five days fermentation at 80-85°F, with three days extended maceration. 100% malolactic fermentation. Gentle pump overs three rackings and lightly filtered. 12 months in older Oak Barrels.

Related information:
Alcohol: 14.5%   **pH**: 3.62   **TA**: 6.45   **Residual sugar**: Dry
Harvest Date: October 12- 15, 1998
Bottling Date: December 1999
Production: 4,615 cases sold at winery and distributed.

Winemaker's Notes:
Ruby red in color, this wine displays an elegant, perfumed nose of blackberry jam, spice with hints of vanilla and wildflowers and a vibrant, jammy, well-structured mid palate. The acid is beautifully balanced with subtle hints of oak that lead to an expressive finish. This lively, yet graceful wine is a beautiful expression of a Mediterranean-style wine.

1998 Palisade Vineyard, Stags Leap

Appellation: Stags Leap District
Vineyard: Palisade Vineyard
Composition: 100% Zinfandel
Vinification and Maturation: Seven days fermentation at 80-85°F, with three days extended maceration. 100% malolactic fermentation. Gentle pump-overs three rackings and lightly filtered. 12 months in one year-old Oak Barrels.

Related information:
Alcohol: 14.5%   **pH**: 3.67   **TA**: 6.36   **Residual sugar**: Dry
Harvest Date: October 1998
Bottling Date: December 1999
Production: 210 cases sold at winery and distributed.

Winemaker's Notes:
Deep ruby in color with hints of purple, our 1998 Zinfandel was created exclusively from our own Palisade Vineyard, located in the Stags Leap District. Loaded with white and black pepper aromas, and well-structured with great

acidity and rich, jammy fruit and spice, this wine is well-anchored in the mid-palate with fresh flavors lasting throughout.

Clos Du Val also produces Cabernet Sauvignon, Reserve Cabernet Sauvignon, Merlot, Pinot Noir, Sangiovese, Chardonnay and Ariadne, a blend of Semillon and Sauvignon Blanc. Clos Du Val also produced a 1994 Zinfandel from estate grapes in the Stags Leap District, released in 1996.

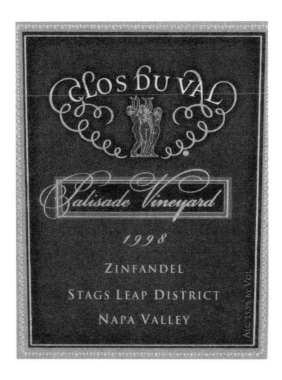

Clos LaChance Wines

Clos LaChance Wines
21511 Saratoga Heights Drive
Saratoga, CA 95070
Phone: 408-741-1796 Fax: 408-741-1198
Email: cherylmurp@aol.com
Website: www.clos.com

Clos LaChance Wines, founded in 1992 by Bill and Brenda Murphy, is a small, family-owned winery, located in the heart of the Santa Cruz Mountains. A small producer, Clos LaChance focuses on hand-made wines from local vineyards and other premium appellations throughout Northern California. The name Clos LaChance is derived from the French word "clos," for the fenced-in area around a vineyard (in this case the owners' home vineyard) and from co-owner Brenda Murphy's maiden name, LaChance.

The Clos LaChance trademark - a hummingbird among the grapevines - is the primary feature on the label. The hummingbird was chosen not only for its grace and style, but also for its strength. The hummingbird is a welcome addition into vineyards for its "energetic" efforts in keeping other birds away.

Owners: Bill and Brenda Murphy   **Winemaker**: Jeff Ritchey

Winemaking Philosophy:
To create a balanced wine that expresses the flavors found in the vineyard, Clos LaChance winemaker Jeff Ritchey strove to make a wine that expresses the true character of the zinfandel grape and the Twin Rivers Vineyard.

1998 Clos LaChance El Dorado Zinfandel

Appellation: El Dorado
Vineyard: Twin Rivers
Composition: 100% Zinfandel
Vinification and Maturation: Cold soaked for two days prior to fermentation. El Dorado produces wines with great intensity and tannins, so Jeff and his crew pressed the grapes at 5 degrees Brix to preserve the fruit and ensure that the tannins were not over-extracted. Upon completion of fermentation, the wine was racked down into barrels.

Related information:
Alcohol: 14.9%   **Residual sugar**: Dry
Harvest Date: October 17th, 1998
Bottling Date: March 20th, 2000
Release Date: June, 2000
Production: 1566 cases
Retail: $17.00

Winemaker's Notes:

The El Dorado Zin is true to the zinfandel grape. Classic raspberry and cherry fruit are accentuated with blueberry notes from the vineyard. A pepper nose, typical of the Zinfandel varietal, is balanced with rich vanilla and spice from the American oak. To the taste, this wine is a rich and fruity, lush and juicy. The finish lingers just long enough to make the drinker want to take another sip.

Cellar Notes:

1998 is the second Twin Rivers Zinfandel from Clos LaChance. It is more true to character than the 1997, with more finesse and power. The 1999 vintage (currently in the barrel) promises to be the best yet. Twin Rivers dropped a significant amount of fruit in 1999 to guarantee ripeness and flavor. At this point, it looks like it has definitely paid off.

Clos LaChance, El Dorado   56

Collier Falls

Collier Falls
9931 West Dry Creek Road
Healdsburg, CA 95448
Phone: 707-433-7373 Fax: 707-433-7780
e-mail wine@collierfalls.com   web site: www.collierfalls.com

Collier Falls Vineyard is a 100-acre hillside property, located in the northwest corner of Sonoma County. The property was purchased in 1997 from Ferrari-Carano Winery by Barry and Susan Collier, and their sons Adam and Joshua. The first vineyard was planted on the property in 1982 by the Meeker Winery. It was comprised of three separate hillside-terraced blocks planted to "old-clone" Zinfandel on St. George rootstock. Vines from this original vineyard still produce approximately three tons per acre. In 1997, Collier Falls added eight additional acres of Zinfandel, and in 1998, ten more acres were planted; three to Zinfandel and seven to Cabernet Sauvignon, Cabernet Franc, Petite Syrah and Petit Verdot.

No tasting room. Mailing list.

Proprietors: The Collier Family   **Winemaker**: Alex MacGregor

1998 Collier Falls Private Reserve Zinfandel

Appellation: Dry Creek Valley **Vineyard Designation**: Collier Falls
Composition: 100% Zinfandel
Vinification and Maturation: Open top punch down three times daily. Closed top, aeratively pumped over three times daily. 15 months in French and American Oak, 30% new. Racked four times previous to bottling. Unfined and unfiltered.

Related information:

Alcohol: 14.5%	**Residual sugar**: Dry
Brix at Harvest:24.5	**Harvest Date**: October 9, 1998
Bottling Date: 2/16/2000	**Release Date**: April, 2000
pH: 3.62	**TA**: 6.3

Production: 525 cases distributed and sold through the mailing list.
Retail: $24.00

Winemaker's Notes:

This wine shows subtle aromas of raspberries and strawberries, followed by spice and vanillin oak components. On the palate, the fruit is very forward, tannins are soft and integrated, and the acid is bright and balanced. This Zinfandel, with its creamy mouth feel, is a pleasure to drink young while the fruit is still vibrant, but oak again and structure ensure added complexity with several years of bottle age.

Collier Falls also produces Cabernet Sauvignon, Cabernet Franc, Petite Verdot, and Malbec.

Cosentino Winery

Cosentino Winery
7415 St. Helena Highway
P.O. Box 2818
Yountville, CA 94599
Phone: 707-944-1220 Fax: 707-944-1254 www.cosentinowinery.com

Cosentino Winery, recognized for its hand-crafted, small production lots, became the first winery to get label approval for the use of "punched cap fermentation" in 1995. In a world where modern technology often wins out, the winery continues to employ this old-style Burgundian method, which requires the manual punching, or mixing, of the skins with the juice.

Tasting room open 10:00 to 5:00 daily.

Proprietor and Winemaker: Mitch Cosentino

Winemaking Philosophy:

The labor-intensive punched cap fermentation deviates from today's more conventional pump-over method, where fermentation takes place in a stainless steel tank and juice is pumped over the floating cap of skins. Hand punching is also space-intensive, taking up to eight times more floor space.

Cosentino has discovered through experimentation that punched cap fermentation is superior to the pump-over method in two important ways. First, it allows for better management of the tannins, resulting in softer, richer tannins. Second, it provides wine that shows more fruit concentration. In 1998 rotary tank fermentation is the preferred method for one key lot of Old Vine Zinfandel.

Mitch Cosentino makes many different wines. Each blend determines the case production which as a result, varies from year to year. While this approach makes forecasting wines difficult, Cosentino wouldn't have it any other way. "The goal is to make the best wine, not the best wine to fit a certain number of cases," he says. "I want the art developing the business, not the business creating the art."

1998 The Zin

Vineyard: The cornerstone of this wine's pedigree emanates from Sonoma County's Russian River Valley, complemented by one Old Vine Vineyard in the Lodi region and selected lots in Napa Valley.

Vinification and Maturation: The red grapes are hand-picked and crushed directly into small, half-ton plastic bins, where they are inoculated with yeast to begin the fermentation process. During the fermentation cycle, which typically lasts from six to ten days, the natural grape sugar is converted to alcohol. Three or four times daily during this period, the cap is manually submerged, pushing the floating skins downward into the juice. Barrel aged for 10 months in both French and American oak.

Related information:

Alcohol: 13.8%   **Production**: 1,800 cases **Retail**: $25.00

W i n e m a k e r ' s   N o t e s :   "The Zin" may sound presumptuous, but it was never intended that way. It began with the 1990 vintage that produced a mere 100 cases of Zinfandel and was casually referred to as "The Zin." Pepper and black fruit dominate this wine.

1998 Cigarzin

Vineyard: A blend of Sonoma and Lodi, with the principal source being a 55 year old head trained vineyard in Lodi.

Vinification and Maturation: The fruit was mostly punched cap fermented in small bins. Unfined. Aged in French and American oak.

Related information:

Alcohol: 13.8%   **Release Date**: in release **Production**: 2200 cases

Retail: $22.00

W i n e m a k e r ' s   N o t e s :   Cigarzin offers a medium weight, dry alternative to the red late harvest and dessert wines most people consider as accompaniments to cigars. It's not as heavy or sweet as traditional "cigar wines." Cigarzin offers more red fruit flavor, as opposed to the black fruit essence typically found in our regular bottling of The Zin.

1998 The Zin Reserve

Vineyard: Russian River

Vinification and Maturation: The fruit was mostly punched cap fermented in small bins. Unfined. Aged in French and American oak.

Related information:

Alcohol: 15.7%   **Production**: 180 cases **Retail**: $50.00

W i n e m a k e r ' s   N o t e s :   The Zin Reserve is a huge intense wine dominated by pepper and black fruit.

Cosentino Winery also produces Red and White Meritage, Chardonnay, Cabernet Sauvignon, Sangiovese, Nebbiolo, Merlot, Cabernet Franc and Pinot Noir.

Dickerson Vineyard

Dickerson Vineyard
86 Lagoon Road
Belvedere, CA 94920
Fax: 415-435-7019

"If it's wine it should be red.
　　If it's red it should be Dickerson"

　　William Dickerson divides his time practicing psychiatry in Marin, teaching medical students in San Francisco, and growing award winning Napa Valley Zinfandel, Merlot, and Ruby Cabernet for Ravenswood Winery and his own Dickerson Vineyard label. An Associate Professor of Psychiatry, U of C, San Francisco, Dickerson taught classes and conducted wine tastings for The Napa Valley Wine Appreciation Course for 20 years. His internationally published wine writings range from vintages of Chateau Latour and the vineyards of Germany to "The Medical & Therapeutic Values of Wine" in The University Of California/Sotheby Book of California Wines. He has been a Napa Valley wine grower since 1971, his wines and vineyard have been highly acclaimed and passionately sought after.

No tasting room. Mailing list.

Owner and Grape Grower: William Dickerson
Winemaker: Joel Peterson

Winegrower Philosophy:
　　To faithfully reflect the vineyard terroir in single vineyard wines:
　　　　Work with nature
　　　　Respect tradition
　　　　Attentive organic care

An ancient oak tree still growing beside our vineyard was the upright of a beam press used in early Napa Valley wine making. This form of wine press was invented for Egyptian pharaohs and has changed little in over 3000 years of continual use around the world before its arrival in our Napa Valley vineyard. To honor the universality and tradition of wine throughout cultural history, this oak is etched on every bottle of Limited Reserve.

1997 Limited Reserve

Napa Valley Zinfandel

Appellation: Napa   **Vineyard Designation**: Dickerson
This 21 acre West Zinfandel Lane Vineyard, bordering Rutherford and St. Helena, is small enough to allow traditional hand care and meticulous attention to organic wine growing. Possible for a small artisinal vineyard, the grapes are dry farmed, organically cultivated.
Vinification and Maturation: Vinified on naturally present vineyard yeasts, 7 - 10 days open top fermentation, punch down. Aged in French oak barrels for 22 to 23 months.

Related information:
Alcohol: 14.5%   **Residual sugar**: Dry
pH:3.03   **TA**: 6.22
Bottling Date: 7/26/99   **Release Date**: December 1999
Production: 205 cases and 90 magnums
Retail: $24.50 per 750ml. $65.00 per Mag. Direct mail

Winemaker's Notes:
The big spicy presentation is an elegant and classic wine proclaiming a great vintage year. The wines incomparable spicy briary fruit, its depth and profound strength without the crude clumsiness from warmer areas proclaim Napa Valley Zinfandel! Its elegance, finesse, and signature minty background identify the terroir and proclaim Dickerson Vineyard! The rich taste, full extract and classic aftertaste of balanced tannin and acidity mark this with the quintessential of Zinfandel complexity and subtlety. Excellent aging potential, long lived.

1998 Limited Reserve Napa Valley Zinfandel

Still aging in small French cooperage, the wine is developing well with brilliant color and a powerful forward bouquet. Typical richness in the mouth and after taste comprise an early offering of fine quality. The wine will be available near the end of the year, again by direct mail.

Dickerson Vineyard also produces Merlot, and Ruby Cabernet, a unique meritage of Cabernet Sauvignon, Cabernet Franc, and Napa Valley's only Ruby Cabernet.

Dover Canyon Winery

Dover Canyon Winery
4520 Vineyard Drive
Paso Robles, CA 93446
Phone: 805-237-0101 Fax: 805-237-9191
Email: dovercanyon@tscn.net
Website: www.dovercanyon.com

Dan Panico has been making wine in Paso Robes for 10 years, first as a lab technician and cellar rat at a large winery, than as an assistant winemaker and later winemaker for Gary Eberle of Eberle Winery, also in Paso Robles. While at Eberle, Dan established himself as a winemaker with a sensitive palate and serious talent. Dan started producing his own wine in 1992, while he was still winemaker at Eberle Winery.

Dan's first vintages were produced from fruit acquired from Norman Vineyards, a well-known and highly acclaimed Westside Vineyard, and his focus is on westside vineyards, particularly dry-farmed, head-trained Zinfandel. His label, Dover Canyon, is named for a canyon in the northwest corner of our appellation.

A quiet, soft-spoken man, Dan offers his version of effusive praise for westside Paso Robles: "Mountain grown fruit. Shale. Limestone. Non-irrigated. Six miles from the Pacific." What this means to the rest of us is that westside Paso Robles, which is hilly and has higher elevations than the eastside, has hard-working, stressed vines producing small-berried, intense fruit with thick skins and dark pigment. Stiff ocean breezes keep the mountainous vineyards free of fog and mildew, and keeps the vines cool during long summer days. A strip of pre-calcareous shale, Dan's favorite soil profile, runs to the northwest, and some of Paso's best and most award-winning vineyards are located on this strip.

At Dover Canyon, all vineyard lots are hand-crafted, and fermented in traditional Burgundian style in small half-ton fermenters. During fermentation red wines are punched down by hand frequently each day. White wines are barrel fermented, and both red and white wines are racked using gravity (for gentle movement with limited air contact). Red wines are unfiltered

"I like to produce premium wines without the bruising caused by pumping and mechanical processing," Dan explains. Although harvest is a labor-intensive time of year for winemakers, Dan feels the real work happens in the vineyard.

"Much of my focus is on acquiring great fruit from vineyard owners who spend a considerable amount of time in the vineyard. I can't stress enough the importance of quality fruit."
Tasting room open to the public Friday through Sunday 11-5. No charge for tasting. No tours.
Owner and Winemaker: Dan Panico

V i n t a g e   N o t e s :
Although 1998 was a tough-to-mediocre year for many producers, one of the advantages of being a very small winery with limited production is that we are not tied to production contracts with vineyards. This allowed us to go "window shopping" in 1998. We selected only the very best fruit from viticulturists who were willing to pick at precisely the Brix and pH levels that we desired. The resulting wines, after two cool, rainy years, offer unusually intense brier-pepper flavors from the old vine vineyards we selected. Although future warmer vintages may bring a return to the jam-and-spice styles we also love, we feel that one positive aspect of the '98 vintage was the eclectic pepper flavors found in old, traditional head-trained vineyards.

1998 Cujo Zinfandel, Paso Robles

Our '98 "Cujo" is a blend of zinfandels from three head-trained, old vine vineyards, Cobble Creek, Swiss Collina, and Dusi, all picked at over 26° Brix. Entirely fermented in small open top bins and aged in neutral French oak for ten months. This Zinfandel will display the intense blackberry and pepper characteristics of head-trained vineyards, with subtle notes of leather, cassis and spice.
Retail: $18.00

1998 Zinfandel Reserve, Paso Robles

Our 1998 Zinfandel Reserve is a showcase of fruit from carefully selected vineyards. Intense, dark and spicy with briary, wild berry flavors. After aging for 16 months in a combination of new and neutral American oak, this wonderful blend of old and new vineyards exhibits spicy pepper, wild berries, sage, smoke and toast.
Retail: $22.00

1998 The Barbarian, Paso Robles

A blend of Zinfandel and Petite Sirah from mature vines, this wine is intense, dark and spicy, with brambly, wild berry flavors that only come from old vine fruit. After aging for 16 months in new French and American oak, this wine exhibits smoky, spicy aromas and luscious wild berry flavors with a touch of mint.
Retail: $16.00

P r o d u c t i o n   N o t e s:
Our professional commitment continues to be production of small, hand-crafted lots of wine, with a focus on premium Paso Robles vineyards. Dan produces several Zinfandels each year, as well as Rhône-style wines and limited productions of Bordeaux varieties.

Dan enjoys working with white Rhône varieties, including Viognier, Roussanne and Marsanne. "We can achieve high fruit aromas and flavors in dry wines. I enjoy interesting whites. Our customer enjoy them too. Our Rhône releases are a refreshing alternative to the few white varietals most people know."

Dan also points out that the chalky, rocky soils of Paso Robles, along with hot day and cool nights make this area ideal for growing Rhône varietals, both white and red. "This is an excellent region for Rhônes."

Dover Canyon also produces Viognier, Roussanne, Syrah and Bordeaux blends.

Dry Creek Vineyard

Dry Creek Vineyard
3770 Lambert Bridge Road
Post Office Box T
Healdsburg, CA 95448
Phone: 707-433-1000 or 1-800-864-WINE Fax: 707-433-5329
dcv@drycreekvineyard.com

Celebrating a quarter century of winemaking, Dry Creek Vineyard was founded in 1972 by David Stare and has grown from a 1,300 case winery to approximately 120,000 cases today. Located in the heart of Dry Creek Valley, Sonoma County, the winery owns about 200 acres of vineyards. These vineyards supply about 25% of the winery's total needs, with the remainder being purchased on long-term contracts from 63 quality-oriented local growers. Each lot of grapes, regardless of origin, is picked, fermented and barrel-aged separately. Out of these lots the best are steered towards the winery's "Reserve" program and are given additional aging and cellar treatment. Spirited nautical packaging and intensive, small-lot winemaking characterize the Dry Creek Vineyard brand.

Tasting Room open daily from 10:30 am to 4:30 pm.

Proprietor: David S. Stare   **Winemaker**: Jeff McBride
Larry Levin was the winemaker for the 1998 vintage.

Winemaking Philosophy:

All Zins are not created equal. Grapes harvested from older vines typically lead the pack qualitatively. That is why at Dry Creek Vineyard, fruit harvested from old vines - ranging in age from 80 to 100 years old - are always our first choice.

One reason Zinfandel from old vines usually produce such great wines is that they were typically interplanted with Petite Sirah, Carignane, and other varieties. The influence of these blending grapes really makes a difference in the finished wine. Zinfandel, on its own, is usually rich in raspberry character, lighter in color, and lacking firm structure compared to other heavier-bodied reds. That is why, for firm structure and depth of color, we blend our Zinfandel with Petite Sirah. Depending on the vintage, we add 10 to 25 percent Petite Sirah to achieve a rounder, fuller, and richer character.

One of the biggest challenges in making Zinfandel is the fact that this varietal ripens so unevenly. Raisined and over-ripe berries hang side by side with immature pink berries on the same vine, or even on a single cluster. The pink berries are tart and not very flavorful, while the raisined ones offer a jammy rich character along with an abundance of sugar. We try to pick when there are nearly no pink berries to achieve a more flavorful wine, one with an appealing hint of jam and a moderately higher alcohol from the sugar concentrated in the raisins.

The Zinfandel we select is typically, but not always, from hillsides. Our Reserve Zinfandel comes from a flat bench area. Some of the newer plantings propagated from the old vine clones also show as good, and in some cases, better quality.

1998 "Old Vines" Zinfandel

Appellation: Sonoma County (66% Dry Creek Valley, 24% Russian River Valley, 5% Sonoma Valley, 4% Alexander Valley and 1% Mendocino County)
Composition: 85% Zinfandel and 15% Petite Sirah
Vinification: Fermented 10 days at 79° F.
Maturation: 13 months in American and French oak.

Related information:
Alcohol: 14.2%   **Residual sugar**: Dry
Harvest Date: October 1-7, 1998
Production: 6,000 cases sold at winery and distributed.
Retail: $18.00

Winemaker's Notes:
Jammy essence of blackberry and raspberry aromas complement briary, smoky, and spicy notes. A fruit forward wine, flavors of raspberry, blackberry, black cherry, and plum meld with rich chocolate and toffee. Its full textured core reveals a firm structure, soft tannins, and creamy oak. A well-integrated Zinfandel, the finish is lush and elegant. Aging potential 5 to 8 years.

1998 Heritage Clone Zinfandel

Appellation: Sonoma County
Composition: 75% Zinfandel and 25% Petite Sirah
Vinification: Fermented 10 days at 79° F.
Maturation: 14 months in American and French oak.

Related information:
Alcohol: 14.5%   **Residual sugar**: Dry
Harvest Date: September 29 - October 26, 1998

Production: 5600 cases sold at winery and distributed.
Retail: $15.00

Winemaker's Notes:

Luscious raspberry and cherry aromas precede intriguing baking spice notes. Juicy, concentrated berry components blossom into textured flavors, revealing a slight earthy quality and dusting of mind. This 1998 Heritage Clone Zinfandel is vibrant and mouth-filling, finishing with a balance of oak and fruit.

Dry Creek Vineyard also produces Dry Chenin Blanc, Fumé Blanc, Reserve Fumé Blanc, Chardonnay, Meritage (red Bordeaux-style), Merlot, Reserve Merlot, Cabernet Sauvignon, Reserve Cabernet Sauvignon.

Please contact the winery for information on their many events. The festivities at Dry Creek Vineyard are well worth scheduling your next wine country visit around.

Easton

Easton
P.O. Box 41
Fiddletown, CA 95629-0041
Tasting Room is located at 10801 Dickson Road, Plymouth.
Phone: 209-245-3117 Fax: 209-245-5415 Winery: 209-245-4277
Email: Terouge@volacano.net

Easton wines are hand-crafted in the Sierra Foothills of Amador County at Domaine de la Terre Rouge, the home for both Easton and Terre Rouge wines. Amador County, located 40 miles southeast of Sacramento, is one of the oldest grape growing areas of California with vines dating back to the 1860's. The Sierra Foothills' granite soils and moderate climate provide the perfect growing conditions for premium grape varietals.

Domaine de la Terre Rouge was founded in 1984 by husband and wife team, Bill Easton and Jane O'Riordan. As one of the founding members of the Rhône Rangers, Terre Rouge was the first to pioneer Rhône varietals in the Amador County. Grenache, Syrah, Mourvèdre, and Cinsault were the first to be planted, followed later by the white Rhône varieties of Viognier, Marsanne, and Roussanne.

While pioneering the new frontier in Amador County, Bill was drawn to the traditional old head-trained vines of the area, and began making artisan zin in 1991 under the Easton label. In 1995 the Easton product line was expanded and now includes several vineyard designated Zinfandels, Barbera, Sauvignon Blanc/Sémillon, Merlot and Cabernet Sauvignon.

Tasting Room open Friday, Saturday and Sunday from 11:00 to 4:00. Mail order and newsletter.
Proprietors: William Easton and Jane O'Riordan
Winemaker: William Easton

Winemaking Philosophy: Bill's focus as winemaker is on the production of a limited amount of high quality wine in the European estate tradition. The wines are hand-crafted in small lots, fermented and aged in French cooperage, with careful attention to detail and as little manipulation as possible. Bill's 26 years in the wine trade has defined Easton style: elegant, rich, and well-balanced wines that compliment a wide variety of foods.

Easton Zinfandel Shenandoah Valley

1996 Vintage

1996 was a special Zinfandel year for us. Our Shenandoah Valley appellation Zinfandel made from old vines grapes ripened to perfection. The flavors this year are more black fruit orientated: wild blackberry, black cherry, spiced plum. Aromas are peppery and complex, evoking other spices. The wines flavors are lush with lots of finesse. Fruit was harvested a degree higher than in 1995 at an average of 25° Brix. 3.4 pH, 7.0 grams per liter total acidity. Our Easton Zinfandels are developing a reputation across the country for power with finesse. By using French oak and judicious production and cellaring techniques we have reined in the aggressive tannins that have sometimes marred Amador Zinfandels in the past.

Related information:
Alcohol: 14.5%   **Brix at Harvest**: 25.0
Harvest Date: 10/07/96   **Bottling Date**: 5/1/98
Release Date: 6/01/98
Production: 1500 cases sold at winery and distributed.
Retail: $18.00

1997 Vintage

The 1997 vintage gave us fruit forward wine with great mouthfeel and freshness. It is delicious to drink now, as a young wine, and should age nicely for about five years. Our Shenandoah Valley appellation bottling is always more forward and expressive with raspberry/blackberry/black cherry fruit flavors, a subtle undercurrent of pepperiness, broad texture, nice balance, and finish. Not overwrought and muddled by excessive wood like some Zinfandels are currently made. Aged exclusively in French Burgundy barrels for 18 months. This wine holds to the Easton Zinfandel axiom of power and finesse.

Related information:
Alcohol: 14.5%   **Brix at Harvest**: 25.0
Harvest Date: 9/2-15/97   **Bottling Date**: 5/1/99
Release Date: 2/01/00
Production: 1500 cases sold at winery and distributed.
Retail: $20.00

1998 Vintage

Although harvest was as late in 1998 as it was early in 1997, it was a tour-de-force year for Easton Zinfandels in 1998. It was a very late set for all grapes in 1998. Fortunately we didn't get the rain that other parts of California did until the latter part of November. We started picking for this bottling on 9/25/98. Grapes were high in sugar and malic acid. Flavors were intense and spicy. Colors were dark. We left the wine in barrel a couple of extra months to round out further.

Related information:
 Alcohol: 14.5%   **Brix at Harvest**: 24.5 to 25.0
Harvest Date: 9/25/98-10/07/98
Bottling Date: 6/28/00   **Release Date**: projected 03/03/01
Production: 2000 cases sold at winery and distributed.

Easton Zinfandel Fiddletown

1996 Vintage

Our 1996 Zinfandel is from the historic Eschen Vineyards, which was originally known as Cox Vineyard and planted sometime prior to the 1869 township survey of Fiddletown according to research done by local viticultural historian, Eric J. Costa. This is the oldest known Zinfandel vineyard in the world with the original four acres planted around 1865 by Cox. Some of our fruit comes from the oldest section of the vineyard that is on an east-facing hillside and bears a meager ½ ton per acre. Overall the vineyard averages about two tons per acre of flavorful fruit. The vines are dry farmed or totally dependent on rainfall and farmed organically. The soil is well drained Sierra series decomposed granite.

Our Fiddletown Zinfandel is always in short supply. Unfortunately, there is never enough to go around…this is old vine wine. This is one of the most misunderstood vineyard sites in California. Even with managed wine production techniques Eschen Zinfandels are highly-structured wines with very fine tannins. They tend to be very closed-in for their first couple of years of life. They are extremely age-worthy and really reward the patient cellarer. Harvest in 1996 was on October 14th with 25° Brix, 3.28 pH, and 6.6 grams total acidity. Flavors are wild blackberry with subtle notes of pepper and fennel. Mouth-feel and balance are exceptional.

Related information:

Alcohol: 14.5%   **Brix at Harvest**: 25.0
Harvest Date: 10/07/96
Bottling Date: 5/01/98   **Release Date**: 6/01/98
Production: 350 cases sold at winery and distributed.
Retail: $20.00

1997 Vintage

In my book, Fiddletown Zinfandel is Eschen Vineyard. There are other Zinfandel vineyards in the Fiddletown appellation, but none as old and distinctive as this one. Gnarled old vines struggle to survive after 100 years of existence with a few vines dating back to 1865. Yields are low and diminishing but in 1997 more of the fruit was made available to us. Fruit was harvested between 24-25° Brix, 3.45 pH, and 7.7 grams total acidity on September 25th and 26th, a full two weeks earlier than 1996. Typical Eschen blackberry fruit with notes of fennel and pepper spice. Great mouth-feel and length.

Related information:
Alcohol: 14.5%   **Brix at Harvest**: 24.7
Bottling Date: 5/01/98   **Release Date**: 06/01/99
Production: 700 cases sold at winery and distributed.
Retail: $20.00

1998 Vintage

Complex, spicy, and wild. Wild blackberries, that is, in aromas and flavors. Great concentration of fruit and a whopping acidity of 7.8/grams per liter to match and balance the fruit. Each year we are allowed to play with a little history. These grapes are like working with great Pinot Noir and that is exactly how we make it. Open-top fermentation, hand punched, aged in Burgundian cooperage, and unfiltered.

Related information:
Alcohol: 14.5%   **Brix at Harvest**: 24.5
Harvest Date: 10/24/98
Bottling Date: 6/28/00   **Release Date**: projected 10/01/00
Production: 300 cases sold at winery and distributed.

Easton Zinfandel Shenandoah Valley

Easton, Amador County

Estate Bottled

1998 Vintage

Our first estate Zinfandel comes from the old vine Baldinelli property directly across the road from our winery. Planted in 1972 on St. George rootstock, it is head-trained and non-irrigated. Careful cultivation, shoot thinning, and the naturally devigorating mountain soils allow us to make spectacular wines from the site. Concentrated black fruit flavors and peppery spices with hints of chocolate and dried flowers. Extra barrel time as with our other bottling, a bit more brand new French oak and bottled in our new estate package. Full-tilt boogie!

Related information:
Alcohol: 14.9%   **Brix at Harvest**: 25.5
Harvest Date: 10/18/98
Bottling Date: 6/28/00   **Release Date**: 10/01/00
Production: 800 cases sold at winery and distributed.
Retail: $28.00

Easton Zinfandel Amador County

1997 Vintage

This is the first time we have offered an entry level zinfandel with out Easton line-up. It is a popularly priced wine that possesses exuberant blackberry fruitiness, a lush, silky texture, and a long creamy finish nuance by 10 months aging in François Frères cooperage. It also has attractive/peppery components.

Related information:
Alcohol: 14.5%   **Brix at Harvest**: 24.0
Harvest Date: 09/02-30/97
Bottling Date: 09/10/98   **Release Date**: 10/01/98
Production: 2000 cases sold at winery and distributed.
Retail: $14.00

1998 Vintage

This is our entry level Zinfandel that is released in the Fall following the previous year's harvest. It is youthful and vivacious with blackberry fruit, spice and pepper components. It drinks and ages like cru Beaujolais from an estate that makes rich, red wine.

Related information:
Alcohol: 14.5%   **Brix at Harvest**: 24.0
Harvest Date: 10/02-30/98
Bottling Date: 9/07/99   **Release Date**: 10/01/99
Production: 3800 cases sold at winery and distributed.
Retail: $14.00

Domaine de la Terre Rouge also produces white and red Rhône-style blends: Vin Gris d'Amador, Terre Rouge Blanc, Enigma, Tête-a-Tête, Terre Rouge Noir. Varietal wines are Viognier, Mourvèdre, Syrah and Muscat à Petits Grains. Under the label Easton production includes an Easton Criolla and Easton Barbera.

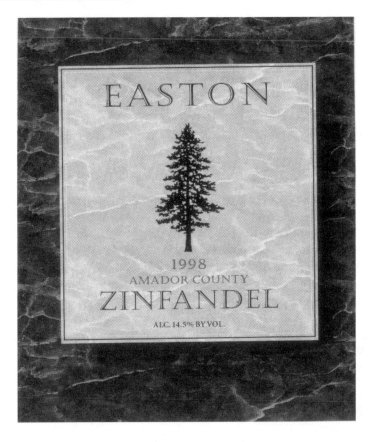

Eberle Winery

Eberle Winery
P.O. Box 2459
Paso Robles, CA 93447-2459
Phone: 805-238-9607 Fax: 805-237-0344
Email: mrviognier@aol.com
www.eberlewinery.com

For over 20 years Gary Eberle has shown a deftness for crafting some of the finest wines in California. Estate grown Cabernet Sauvignon has been Eberle's flagship wine since the winery opened in 1984. Tours of Eberle's wine caves provide a unique experience to view over 16,000 square feet of underground tunnels designed to house traditional oak barrels for the aging of red wines. A hospitality room is located in the caves, which can seat up to eighty people for special wine tasting and dinners. Located off the tasting room is a picnic patio with panoramic view of estate vineyards, available for Eberle guests. Tasting Room and picnic area open daily from 10:00 to 5:00 September through May, and 10:00 to 6:00 June through August. Mail order. Wine Club.

Proprietor: W. Gary Eberle   **Winemaker**: Bill Sheffer

Winemaking Philosophy: 100% Paso Robles ✍ Best varieties for our Appellation ✍ Best vineyards for these varieties ✍ Maximum expression of the quality of fruit with minimal winemaker interference.

1997 Sauret Vineyard Zinfandel

Appellation: Paso Robles
Vintage: An early spring started the vines two weeks earlier than normal. Cooler than normal summer pushed the ripening two weeks later. This combination gave the vines a month more growing time which in turn produced one of the best crops in the past quarter century.
Vineyard Designation: Sauret Vineyard
Composition: 100% Zinfandel
Vinification and Maturation: The grapes were hand picked and delivered to the winery within hours. Destemmed and crushed into stainless steel fermenters and allowed to sit twenty four hours before

inoculation. The wine was pumped over twice a day for 12 days, pressed and allowed to complete malolactic fermentation. The wine was aged in American oak barrels for 12 months.

Related information:

Alcohol: 15.5%   **Residual sugar**: 0.2%
pH: 3.24   **TA**: .71
Brix at Harvest: 25.5   **Harvest Date**: 09/03/97
Bottling Date: 02/22/99
Production: 2,507 cases sold at winery and distributed.
Retail: $20.00

W i n e m a k e r ' s   N o t e s :
Of all the wines that I make, the Sauret Zinfandel is the most consistent in quality from year to year. The 1997 vintage exhibits all the characteristics you would expect from a 30 year old, dry farmed vineyard. This is a nicely balanced wine with a lot of raspberry fruit and spice. It has firm acidity and full body which contributes to a long finish.

1997 Steinbeck Vineyard Zinfandel

Appellation: Paso Robles
Vineyard Designation: Steinbeck Vineyard
Composition: 100% Zinfandel
Vinification and Maturation: Same as for Sauret Vineyard.

Related information:

Alcohol: 14.0%   **Residual sugar**: 0.2%
pH: 3.43   **TA**: .65
Brix at Harvest: 24.4   **Harvest Date**: 09/03/97
Bottling Date: 02/23/99
Production: 2,147 cases sold at winery and distributed.
Retail: $16.00

W i n e m a k e r ' s   N o t e s :
The fact that this vintage is of a richer, more intense version of the 1996 vintage indicates to me that this vineyard is maturing nicely. I believe this vineyard to be capable of producing the type of Zinfandel that has come to be expected from the Paso Robles region.

Eberle Winery also produces Cabernet Sauvignon, Chardonnay, Barbera, Syrah, Counoise, Muscat Canelli, Viognier, Sangiovese and a 5-Variety Rhône blend.

Edmeades Estate Winery

Edmeades Estate Winery
5500 Hwy 128
Philo, CA 95466
Phone: 707-544-4000 Fax: 707-544-4013 Winery: 707-895-3232
Telemarketing: 800-769-3649

Edmeades, located in the tranquil Anderson Valley, has established a reputation for crafting rich, full-flavored old vine Zinfandel through minimalist winemaking techniques. The wines are crafted in a way that exemplifies both the personality of the vineyard and the terroir. Early bottlings of the Ciapusci and Zeni Vineyards has been instrumental in the survival of old vine Zinfandel in Mendocino.

Wine Club. Wines can be sample at The California Coast Wine Center. 5007 Fulton Road, Fulton, CA 95439 Tastings from 10-5 daily.

Proprietor: Jess Jackson   **Winemaker**: Van Williamson

Winemaking Philosophy:

Edmeades wines are crafted using minimalist techniques to insure that the vintage, vineyard, and varietal characteristics are evident once the wine is in the bottle. The grapes are allowed to ripen fully, are hand-harvested into small, shallow, one-half ton bins which allow the grapes to be delivered in optimum condition, without any juicing.

Nutrients are not added prior to fermentation. We encourage the ambient, native yeast and bacteria from the vineyard to be part of the fermentation process. We stress the yeasts, to produce a wider range of aromas and allow the fermentation to struggle to help define the vineyard's personality. The wines are fermented in small bins outdoors then pressed directly to barrels where fermentation is completed. We age all our reds sur lie imparting a chocolate creamy flavor on the finish.

We let the vineyards express themselves, emphasizing a sense of place. Assessing each wine independently, we try not to intervene with the naturally occurring processes. Minimal intervention...that defines Edmeades.

1998 Zinfandel
Eaglepoint Vineyard Mendocino

Vineyard: This Zinfandel bears the unmistakable stamp of Eaglepoint Vineyard, a favorite source of Edmeades grapes. The head-trained vines, 1,800 feet above Ukiah Valley, grow in Timber red clay loam and rocky, fractured sandstone. Western slopes ensure ample daytime sun, while steep temperature drops at night bring on slow, even ripening. Edmeades blends in Syrah and Petite Sirah for added complexity, experimenting with varietal blends in search of richness, much like the Italian immigrants who first made wine in these hills in the 1800's.

Composition: 76% Zinfandel 19% Syrah 5% Petite Sirah

Vinification and Maturation: Grapes were hand harvested and sorted into individual vineyard lots, then placed in bins to cold soak to extract dark color from skins. For greater complexity, the juice was fermented using only native yeasts in small one-ton, open-top bins. The caps of the skins were punched down by hand twice a day to soften tannins and maximize color extraction. After 15 days of maceration, the wine was racked into small oak barrels for 16 months of aging. Unfined, unfiltered.

Related information:
Alcohol: 16%
Residual sugar: .02
TA: 0.60g/100ml
pH: 4.11
Bottling Date: February 2000
Production: 401 cases

Winemaker's Notes:

This is a wine that changes a lot in the glass. The aromas are packed solid with raspberry, rose petal, pepper and spice. As the wine sits, it evolves into layers of more exotic aromas - forest floor, chocolate, coffee, mingled with toasted oak. The flavors start off with sour cherries and end with a creamy dark chocolate aftertaste. Exotic berry aromas combined with classic claret aromas make this a unique Zinfandel. Blending with Syrah and Petite Sirah makes it seem more Rhône in style. Drink this wine slowly enough to allow the aromas and flavors to mature. - Van Williamson, Winemaker

1998 Zinfandel
Zeni Vineyard

Vineyard: This single-vineyard, old-vine Zinfandel is emblematic of the rich wine heritage found among Mendocino mountaintops. In a prized vineyard owned for generations by the Zeni family, head-trained vines more than 80 years old grow in Timber red clay loam and rocky, fractured sandstone. Southern slopes ensure ample daytime sun, while steep temperature drops at night bring on slow, even ripening. In the bottle, the intense fruit justify Zeni's reputation as a jewel in the Mendocino Ridge crown.

Composition: 100% Zinfandel

Vinification and Maturation: Grapes were hand harvested and sorted, then placed in bins to cold soak to extract dark color from skins. For greater complexity, the juice was fermented using only native yeast in small one-ton, open-top bins. The caps of the skins were punched down by hand twice a day to soften tannins and maximize color extraction. After 21 days of maceration the wine was racked into small French and American oak barrels for 18 months of aging. Unfined, unfiltered, unstirred.

Related information:
Alcohol: 16.5%
Residual sugar: .08
TA: 0.62g/100ml
pH: 3.83
Bottling Date: February 2000   **Production**: 266 cases.

Winemaker's Notes:
Can flavors be too intense? Can aromas be too thick? Intensity of aromas and flavors is the wine's profile. Fruit flavors grab the palate and shake it up for the first two tastes. Thereafter, the wine's richness builds upon the palate. Smoked meat, cigar box and blackberry flavors finish with a warming sensation. A clinging finish, creamy coffee and chocolate aromas, white chocolate, caramel, like a candy shop. - Van Williamson, Winemaker

1998 Zinfandel
Ciapusci Vineyard Mendocino

Vineyard: Vines more than 100 years old cover the westerly slopes of Ciapusci Vineyard. This compact mountain-top plot tells the story of all the quality that can come from the Mendocino Ridge viticultural area, and has been a mainstay of prize-winning Edmeades Zinfandels for years. The head-trained vines file down steep slopes of red clay loam. Western slopes ensure ample daytime sun, while deep temperature drops at night bring on slow, even ripening. Edmeades quality begins in these mountains, like no other winegrowing region on earth.

Composition: 100% Zinfandel

Vinification and Maturation: Grapes were hand harvested and sorted, then placed in bins to cold soak which extracts dark color from the skins. For greater complexity, the juice was fermented using only native yeasts in small one-ton, open-top bins. The caps of the skins were punched by hand twice a day to soften tannins and maximize color extraction. After 15 days of maceration, the wine was racked into small oak barrels for 16 months of aging. Unfined, unfiltered.

Related information:
Alcohol: 17.5%
Residual sugar: 0.39
TA: .081g/100ml   **pH**: 3.62
Bottling Date: February 2000

Winemaker's Notes:
This is a winter-time celebration wine. Aromas of nutmeg, Christmas fudge, egg nog and smoke from the fireplace. Silky, soft flavors for a wine so high in alcohol. Dense, concentrated flavors with a long finish. The second glass tastes better than the first. Amazing drinkability. A teddy-bear style wine. - Van Williamson, Winemaker

1998 Zinfandel
Mendocino Ridge

Vineyard: Our Mendocino Ridge Zinfandel shows in a single bottle how grapes from venerable high-altitude Mendocino vineyards can meld in a stunning portrait of this North Coast wine region. Edmeades has been closely linked to these coastal highlands for more than 25 years. In celebration of its recent AVA designation, this wine commemorates the unique caliber of Zinfandel grown among these century-old Italian homesteads high above the Pacific Ocean. Unlike other viticultural areas, Mendocino Ridge is non-contiguous, defined in terms of elevation - only vineyards at 1200 feet or higher qualify.

The legendary Zeni, Ciapusci and Alden vineyards provide most of the fruit for this blend. They appear as small parcels dotting ridge-tops rising through coastal fog. The vineyards enjoy enough sun to ripen Zinfandel, while the coastal influence keeps the nights cool and ripening even. The flavor profile of the blend is amazingly broad. Zeni contributes a briary, wild blackberry fruit character. Ciapusci offers notes of white pepper and red plum fruit flavors. Alden adds floral aromatics and sweet raspberry flavor. Fourteen percent of the blend consists of Syrah and Petite Sirah from our Eaglepoint vineyard, high above the warm Ukiah Valley and a mainstay of Edmeades for years.

Composition: 86% Zinfandel 9% Syrah 5% Petite Sirah

Vinification and Maturation: Artisan winemaking techniques optimize fruit flavor and complexity from each of four prized vineyards. Grapes were hand-harvested and sorted into individual lots, then placed in bins to cold soak for three days to extract rich color from the skins. The juice was fermented using only native yeast. The caps of the skins were punched down by hand twice a days, a labor-intensive process used to soften tannins and maximize color extraction. After three weeks of extended maceration, the wine was racked into small oak barrels for 14 months aging. It is bottled unfined and unfiltered.

Related information:
Alcohol: 15.5%
TA: 0.64g/100ml
pH: 3.74
Bottling Date: February 2000   **Production**: 531 cases

Winemaker's Notes:
The delicate citrus blossom and nectarine aromas live just beneath the stronger cranberry and raspberry fruit and spiciness. As the wine opens, vanillin and toast combine with the bright red fruits. The wine has good creamy flavors, but the acidity focuses the cherry flavors on the long finish. - Van Williamson, Winemaker

Edmeades Estate Winery and Vineyards also produces Pinot Noir, Chardonnay and small quantities of Petite Sirah and Gewürztraminer.

Elyse Winery

Elyse Winery
2100 Hoffman Lane
Napa, CA 95449
Phone: 707-944-2900 Fax: 707-945-0301
Email: elysewine@aol.com

Elyse Winery was started in 1987 by Nancy and Ray Coursen. After nine years of custom crushing at various wineries in Napa Valley we have now found a home. Elyse Winery LLC purchased a small winery on Hoffman Lane just south of Yountville in July of 1997. Total production for Elyse Winery is 6,000 cases. As the winery has grown, it has and will continue to use premium vineyards, work closely with its growers, and allow the vineyards and vintage to sculpt its wines.

Owners: Nancy Cuthbertson and Ray Coursen   **Winemaker**: Ray Coursen

Winemaking Philosophy:
Our passion has always been and will always be rich, fruity red wines. Our main focus has been Zinfandels working with old vineyards, and creating blends that tease the palate with an array of aromas and fruit flavors. While making Zinfandel from the Morisoli Vineyard, we have also started producing Rutherford Bench Cabernets. Our newest varietals are Rhône-style blends of which we are making three uniquely different wines.

The Vintage:
The rains stopped slightly earlier in 1998 than in past years, but the warmth of spring never really materialized. This cool weather resulted in a postponement of bud break, a slowing down of shoot growth, and a delay of flower set until early June. In mid-June, we received a series of rains, which further reduced the crop yield.

As with spring, summer never seemed to arrive. The season was marked by unusually temperate weather patterns resulting in continually delayed harvest dates. In August we waited and waited. Potential disaster loomed on the horizon until the arrival of September. With some light rains in September, we were forced to entertain the possibility of an early rainy season and start thinking about possible adjustments in our winemaking styles for this vintage. Thankfully, our hopes for salvation from the rains were realized with the onset of a beautiful Indian summer in mid-September and October. The diminished crop yield and the warm fall left us picking thirty days later than in 1997. All in all however, those of us who waited out the weather were rewarded with another good vintage.

The 1998 vintage is reminiscent of the 1993 and 1996 vintages - characterized by a lean towards a more feminine style with excellent balance and finesse. Now for those of you who might have read that the 1993 and 1996 weren't very good vintages, go down to your cellars and reward yourself. Remember that all great vintage don't have to age for twenty years.

1998 Sierra Foothills Zinfandel

The year 2000 celebrates the inaugural release of our 1998 Zinfandel made from Sierra Foothills fruit. The Zinfandel is characterized by brilliantly focused fruit flavors and aromas, including wild cherry and cassis, spice, pepper and a hint of orange zest.

As is often the case with fruit grown in the hot summer days of Sierra Foothills appellation, the wine is very rich with jammy textures layered with the aforementioned flavors and aromas of fruit and spice. Great separation of flavors and the long, sweet, oak finish make this a Zinfandel that beautifully accompanies exotic cuisine or certainly offers the strength and fortitude allowing it to stand alone.

1998 Howell Mountain Zinfandel

Winemaker's Notes:

Our Howell Mountain Zinfandel is a wine whose deep, dark purple color honestly reflects its intensity and concentration of flavor. The wine is rich and palate-infused by the flavors and aromas of blackberries, cherries, boysenberries, and white pepper with a hint of ruby grapefruit. Its concentration and richness carry the wine's intense fruit flavors across the palate. The toasty, smoky coffee flavored oak extends a lingering finish to the wine.

EOS Estate Winery

EOS Estate Winery
P.O. Box 1179
Paso Robles, CA 93447
Phone: 805-239-2562 Fax: 805-239-2317
website: www.eosvintage.com

A beautiful showcase winery located seven miles East of Paso Robles on Highway 46, surrounded by 1000 acres of vineyards amongst the gently-rolling oak-studded hills. The estate includes a rose garden, tasting room/Mediterranean marketplace, picnic grounds and self-guided tour. Plans for an Italian Villa-style bed & breakfast are underway. Winemaking facilities are state-of-the-art, with a 26,000 square foot, temperature and humidity controlled barrel warehouse and underground dining facility. Events include Winemaker dinners, weddings, concerts, barbecues and craft-shows.

Proprietors: The Arciero, Vix and Underwood Families
Winemaker: Stephen A. Felten

Winemaking Philosophy:
Personal, hands-on attention to detail every step of the process, from pruning, crop level, canopy management and irrigation in the vineyard to selection of proper toast level for barrel aging. EOS Estate Winery uses no formulas, every decision is based on how the wine tastes at a given time. EOS Estate Winery's winemaking team are artisans practicing their craft, not just employees doing their job.

1998 Estate Bottled Paso Robles Zinfandel

Appellation: Paso Robles   **Vineyard Designation**: Peck Ranch
Peck Ranch is located on a gravely bench above the Estrella River Valley, which funnels cool, marine air into the Northeastern corner of San Luis Obispo County. The forty-year-old vines from this vineyard produce highly-sought-after fruit with intense raspberry and spice.
Composition: 88% Zinfandel, 10% Petite Sirah, 2% Cabernet Sauvignon.

Vinification: Warm fermentation temperatures for maximum extraction. Cap gently irrigated three times daily until dry. Pressed and allowed to complete Malolactic Fermentation before barrel aging.

Maturation: 100% aged in French and American oak barrels for 13 months, with 35% new oak with the balance in one-year-old oak. Wine was splashed out and back to barrels several times in the first six months to develop tannin structure.

Related information:

Alcohol: 14.5%	**Residual sugar**: Dry
Brix at Harvest:23.8	**Harvest Date**: 10/19/98
Bottling Date: 3/15/00	**Release Date**: July 2000

Production: 7,000 cases sold at winery and distributed internationally.
Retail: $15.99

Vintage Notes:

1998, an El-Niño year, was one of the finest vintages of the decade. An extremely wet and cool spring, delayed ripening into the late fall, and an Indian summer into November allowed for slow, even development of flavor intensity.

Winemaker's Notes:

A traditional Paso Robles Zinfandel, this wine has intense blackberry and ripe plum aromas, with complexities of anise, mint and vanilla. Layered flavors of brambly fruit, smoky oak and caramel lead to a balanced, well-structured finish with firm acidity and velvety-smooth tannins. Serve with spicy lamb, barbecued steak or, better yet, with a rich, chocolate dessert.

EOS Estate Winery also produces Cabernet Sauvignon, Petite Sirah, Chardonnay, Sauvignon Blanc and Late Harvest Muscat Canneli.

Eric Ross Winery

Eric Ross Winery
P.O. Box 2156
San Anselmo, CA 91979
Phone: 707-874-3046   Fax: 415-332-7141

Eric Ross Winery is located on a ridgetop off Harrison Grade above Occidental in the Russian River Valley with vineyards that date back to 1937. With case production at 2,000, Eric and John plan to remain small and focused. Seventy percent of the grapes used in the Zinfandel are estate grown.
Tasting room open by appointment only. Mail order.
Proprietors and Winemakers: Eric Luse and John Ross Storey

Winemaking Philosophy: With great attention to detail and care of the wines from grape to the bottle brings the extra level of distinction that the winery prides itself on. Eric Ross is dedicated to remaining a small, unique, hands on winery.

1995 Old Vine Zinfandel

Appellation: Russian River Valley
Vineyard Designation: Occidental Vineyard
Grapes for this wine came from a head pruned vineyard dating back 63 yeas. This mountain vineyard, located in Occidental sits at 1200 feet overlooking the Russian River Valley. The vineyard is dry farmed and yields less than 1 ton per acre producing a wine of great concentration.
Composition: Zinfandel with a field blend of Petite Sirah, and Muscat.
Vinification and Maturation: Hand picked. Manuel punch downs twice a day for two weeks in open top fermenters for primary fermentation with malolactic finishing in the barrel. After pressing, the wine was transferred to American and French Oak barrels for 18 months of aging before bottling.

Related information:
Alcohol: 14.6%   **Residual sugar**: 0%
pH: 3.26   **Harvest Date**: October 1997
Bottling Date: May 1, 1998   **TA**: .78 g/100ml
Production: 652 cases sold at winery and distributed.

Eric Ross Winery, Russian River   86

Winemaker's Notes:
Manual punch downs are used to extract the maximum color and fruit while creating a wine with gentle tannins. 1997 marks the third release of Eric-Ross Old Vine Zinfandel. Aromas of black pepper, cedar, cocoa abound with a spiciness found unique to Zinfandel. Opulent richness continues with hints of deep berry overtones that linger on the palate.

Food pairing: Hearty wine for hearty foods. Lamb and Beef.

1997 Zinfandel

Appellation: Russian River Valley
Vineyard Designation: Occidental Vineyard
Composition: 100% Zinfandel
Vinification and Maturation: Using a combination of yeast the grapes were fermented in open top redwood tanks, then punched down by hand twice a day for two weeks After pressing the wine was transferred to American and French oak barrels for 18 months of aging before bottling.

Related information:
Alcohol: 13.5%	**Residual sugar**: Dry
pH: 3.38	**Harvest Date**: October 1997
Bottling Date: May 1, 1998	**TA**: .80 g/100ml

Production: 440 cases sold at winery and distributed.

Winemaker's Notes: 1997 marks the fourth release of our Russian River Valley Zinfandel. Aromas of blackberries and pepper in the nose are followed in the mouth by a rich jammy quality with firm acidity. This yummy wine screams for food to complement it's forward character.

Eric Ross Winery also produces Merlot, Zinfandel Port and Pinot Noir.

F. Teldeschi Winery

F. Teldeschi Winery
3555 Dry Creek Road
Healdsburg, CA 95448
Phone: 707-433-6626
Fax: 707-433-3077

Located in Dry Creek Valley, Frank Teldeschi began farming his vineyard in 1946. He accumulated five ranches before his passing in 1985. It was in that year that the grapes were harvested for the first wines under the Teldeschi label. In 1993 construction was completed on the winery that bears his name. It is dedicated to the memory of Frank Teldeschi: Grapegrower, winemaker and farmer. His son, Dan, continues on with the same passion and dedication to the vineyards and the wine. Annual production is 2,000 cases.

Tasting Room. (Sometimes you have to knock hard on the door as Dan is chief winemaker and barrel washer.) Mail order. Tours by appointment.

Proprietor and Winemaker: Dan Teldeschi

Winemaking Philosophy: I like to produce full-bodied, unfined, unfiltered wines that are rustic, intensely fruity and powerful. Zin is picked at 24 -25° Brix as it needs to be fully ripe to get full flavors. Blending in Petite Sirah adds color, backbone and structure along with a secret variety for another hue of purple and more fruit flavors. I go for maximum extraction during fermentation with temperatures to 85° and aggressive cap management. I want to get all the goodies out of the grape. Aging takes place in neutral barrels to preserve the fruit. My approach is to make high quality, hand crafted wines that will age a minimum of ten years.

The Vintage: The 1998 vintage, in my opinion, is the lightest of the decade. Late to begin with, we fought cool weather in late summer and suffered through early fall rains that began rot in some of the grapes before they became fully mature. As a result, the decision was made to pick before the Brix got to 24.5 to avoid losing the crop to bunch rot. I am happy with my choice. The wine is still in barrel and showing good fruit, some pepper with tannins that are very soft.

The Vineyard: Estate Grown grapes. I use our vineyards, vineyards my father planted. They are old vines - 85 years old - dry farmed, head pruned, with limited yield and produce fruity wines.

Current Releases

1994 Zinfandel Lot 7, Dry Creek Valley

Composition: 75-80% Zinfandel, 5-7% Petite Sirah and 15 - 20% secret variety.

Vinification and Maturation: Fermentation takes place in stainless steel, open top fermenters with temperatures allowed to get to 85°. Cap is punched down by hand frequently. Drawn and pressed around 4° Brix. Malolactic is induced between 5 to 10° Brix. Blended than aged in ten to fifteen year old French and American oak barrels for 16 to 18 months. Bottled then aged 2 to 3 years in bottles before releasing to give the wine a chance to mellow out.

Related information:

Alcohol: 13.8%	**Residual sugar**: Dry
Brix at Harvest: 24-24.5	**Harvest Date**: Late September, 1994
Bottling Date: 09/96	**Release Date**: May 97

Production: 750 cases sold at winery and distributed.
Retail: $20.00

1994 Zinfandel Lot 6, Dry Creek Valley

The 1994 Lot 6 Zinfandel was released in early 1999. It is the heaviest bodied with bright up-front fruit, large mouthfeel and a lingering finish with firm tannins. This wine will continue to get better and probably peak in 6 to 8 years.

1994 Zinfandel Lot 1, Dry Creek Valley

The 1994 Lot 1 Zinfandel has yet to be released. Medium-bodied to full-bodied with good fruit all the way through. Medium tannins. This wine will age for another 6 to 8 years.

1993 Zinfandel Lot 6, Dry Creek Valley

The 1993 Lot 6 Zinfandel, released May 1999, is probably one of the best balanced wines I have ever produced. Balance in terms of acidity, body, fruit and tannin level. While drinking wonderfully in the year 2000, this wine will continue to improve for another 5 to 7 years.

1992 Zinfandel, Dry Creek Valley

Just one lot produced in 1992. This wine is aging beautifully. Heavy-bodied, subtle berry fruit, huge cherry and plum notes, this wine is big in the flavor department with a long lingering finish. As with all my Zins, this is a great food wine.

1995 Zinfandel Lot 1, Dry Creek Valley

The 1995 Lot 1 Zinfandel is fleshier than the '92. It has lots of fruit, especially blackberry and black cherry. I am pouring this wine in the tasting room at this writing, but the wine has not officially been released. This wine will age for another 7 to 10 years. The '95 vintage in respect to Zinfandel is, in my opinion, the best of the decade.

F. Teldeschi also produces Petite Sirah, Terranova (Dry Creek Valley Red Wine), Late Harvest Muscat Frontignan and White Zinfandel.

Fanucchi Vineyards

Fanucchi Vineyards
P.O. Box 159
Fulton, CA 95439
Phone: 707-545-6806 Fax: 707-544-6106

Being a wine grower at the Fanucchi Wood Road Vineyard since 1972, I wanted to have my efforts in the vineyard to show as an individual wine. I started Fanucchi Vineyards in 1992 using selected grapes from the vineyard I work to make my own vineyard designate wine. I still sell many of the grapes I grow, taking pride in the fact that they find their way into wines crafted by other winemakers.

No tasting room. Limited mail order. Vineyard tours in association with the Russian River Valley Wine Growers event, "Grape to Glass" held every August.

Proprietor, Wine Grower and Winemaker: Peter Fanucchi

Winemaking Philosophy:

I grow the wine in the vineyard. My aim each year is to get the best possible fruit from each vine, balancing it with nature and each vines' abilities. My goal for the wine is to showcase the fruit first.

The Vintage:

The 1998 El Niño weather patterns made for a challenging growing season. Each year, for nearly a century, the vines spend their winter asleep in standing water, awakening in spring as the water recedes and the soil dries. But this year, in the lowest spot of the vineyard, pools of water stayed and ducks were actually swimming around the trunks as the vines struggled to produce new canes and leaves. Winter never left and the rain continued into June. Somehow the elderly vines adapted but about ten percent of the old vineyard showed signs of root-rot beginning. Time will tell if these vines recover. We didn't experience a typical spring in 1998 but the vines did bloom in June and while some trees and grapevines in the area died, the Zinfandel battled its way into summer. In July the plants were hit with extreme temperatures and August came and almost went before the grapes even started to color up. The end of summer and fall tortured us with cool weather, but after a few scary days in the beginning of October, the weather warmed enough to slowly ripen the grapes.

1998 Old Vine Zinfandel

Vineyard: The Fanucchi Wood Road Vineyard planted in 1906.
Composition: 85% to 95% Zinfandel. Field blend with Petite Sirah and Alicante Bouschet.
Vinification: Capture the fruit, capture the fruit, capture the fruit.
Maturation: Aged eleven months in a mixture of shaved French and American oak, medium toast.

Related information:
Alcohol: 15.6%   **Residual sugar**: Dry
Harvest Date: 10/21/98
Bottling Date: 11/12/99   **Release Date**: 2001
Production: 383 cases distributed.
Retail: $39.00 - $60.00

Winemaker's Notes: I know that fantastic fruit grows in the Fanucchi Wood Road Vineyard because I, with my own hands, make sure it does.

Fanucchi Vineyards also produces a Trousseau Gris. Fanucchi Wood Road Vineyard has the only Trousseau Gris Vineyard in Sonoma County.

Fife Vineyards

Fife Vineyards
P.O. Box 553
St. Helena, CA 94574
Phone: 707-485-0323 Fax: 707-485-0832
Tasting Room, located at 3620 Road B in Redwood Valley, is open daily from 10:00 to 5:00.
Mail order.
Winemaker: John Buechsenstein

1998 Whaler Vineyard Zinfandel Mendocino

A very special place
The Whaler Vineyard faces the warm afternoon sun on a gentle southwest-facing rocky slope on the Talmage Bench in Mendocino County (just east of the Russian River, a few miles south of the town of Ukiah). This excellent exposure and the well-drained soils allow for ideal ripening of Zinfandel. Old timers in the area will tell you that this is often the first vineyard in the valley to ripen - the vineyard everyone else watches to gauge how they are doing. The massive, old, cordon-trained Zinfandel vines typically generate intense dark berry flavors and 'jamminess' in the wine.

What To Expect From This Wine
For zinfandel, our goal is to create a wine with concentrated fruit that retains a liveliness, beauty, and sense of place. We use very gentle winemaking and try to hold down the alcohol and avoid over ripeness. We seek to produce very complex wines of beauty that continue to develop nuances of flavors over several years of enjoyment. This is a wine that opens quickly to soft, jammy flavors, yielding to black cherry fruit-leather, dense chocolately-char, and mouth-watering, multi-layered fruit depth.

About the Harvest
The 1998 season was heavily influenced by wet El Niño weather into late spring delaying all stages of vine growth and development by up to a month for some varieties. As a consequence the vintage was of below

average size as it produced both fewer and smaller berries. The Whaler Vineyards ease of ripening allowed the fruit to develop richness and concentration in spite of a relatively short and late season. The 1998 Whaler Vineyard was harvested September 18th.

The Gentle Art of Winemaking

Our winemaking philosophy is simple: obtain grapes with personality, handle them minimally, then focus on lending to bring out nuance and character in the wine. The fruit is harvested by hand, and is destemmed with minimal crushing to avoid harsh skin and stem tannins and to preserve the fruit intensity. The must is fermented in small lots, with regular gentle pumpovers. The wine is pressed at dryness and racked to a combination of new and used French oak barrels for twelve months of aging. The 1998 Whaler Zinfandel was bottle in July 1999.

Related information:
Varietal Composition: 96% zinfandel, 4% petite sirah
Alcohol: 14.8%
Total Acidity: 6.5g/L
pH: 3.82
Release Date: February 1st, 2000
Production: 1000 cases sold at winery and distributed.
Retail: $20.00

1998 Old Vines Zinfandel Napa Valley

An Historic Gem

Since the middle of the last century, the area around Larkmead Lane between St. Helena and Calistoga has been know as one of the best places in the Napa Valley to grow red grapes. Our vineyard there is planted with 75-year-old zinfandel in a gravel streak formed by the alluvial fan of Ritchie Creek as it runs down Spring Mountain on its way to join the Napa River. In addition, we purchase Zinfandel from a neighbor's old-vine vineyard that also expresses the personality of the microclimate - exuberant, warm-toned fruit, a robust yet elegant structure, and a lively, flannel-soft texture.

Our goal is to create a wine that expresses it's place and possesses concentrated fruit while retaining a liveliness and beauty. Our winemaking emphasizes gentle handling of the fruit and the avoidance of extreme extraction. We look for fruit ripe enough to fully develop the flavors but without the overripe flavors of raisins. We seek to produce complex wines that continue to develop nuances of flavors over several years of enjoyment.

What To Expect From This Wine

We find this vintage of Our Old Vines Zinfandel to have soft, supple rich fruit suggesting grenadine and black raspberry jam. As is often the case with old vines, the wine has a real purity of expression with every facet of its flavor in focus.

About the Harvest

Napa Valley's typical October "Indian summer" with warm, dry conditions created the ideal perfect finish to an otherwise difficult season. The small berries and long hang time while ripening allowed the fruit to develop richness and concentration is in spite of a relatively short and late season. We completed harvest of the 1998 Old Vines Zinfandel on October 14th.

The Gentle Art of Winemaking

Our winemaking philosophy is simple: obtain grapes with personality, handle them minimally, then focus on lending to bring out nuance and character in the wine. The fruit is harvested by hand, and is destemmed with minimal crushing to avoid harsh skin and stem tannins and to preserve the fruit intensity. The must is fermented in small lots, with regular gentle pumpovers. The wine is pressed at dryness and racked to a combination of new and used French oak barrels for twelve months of aging. The 1998 Zinfandel was bottled in November 1999.

Related information:
Varietal Composition: 100% zinfandel
Alcohol: 13.9%
Total Acidity: 5.8g/L
pH: 3.81
Release Date: February 1st, 2000
Production: 3600 cases sold at winery and distributed.
Retail: $22.50

1998 Mendocino Uplands Zinfandel Mendocino

Fife Vineyards, Redwood Valley   95

The Best of the Hillsides
Upland refers to the many hillside and benchland sites around Lake Mendocino which are home to cherished small blocks of old red vines, many tended by the third generation of their original founders. Vine age and continuity of production are testament to the quality of these fortunate growing sites. To a great extent, these vines are self-limiting in yield and ripen smaller crop loads to mouth-watering intensity. It is exciting to utilize these older sites in a blend that expresses the best of Mendocino fruit.

What To Expect From This Wine
Cranberry-cassis, black raspberry jam...classic spicy Zin-berry fruit. With its soft entry, luscious fruit and spice flavors, an easygoing appeal, our Mendocino Zinfandel blend is a versatile companion to a wide variety of cuisines.

About the Harvest
The Uplands vineyards ease of ripening was a major benefit in this somewhat difficult year. Small berries and long hang time at ripening allowed the fruit to develop richness and concentration in spite of a relatively short and late season.

The Gentle Art of Winemaking
Our winemaking philosophy is simple: obtain grapes with personality, handle them minimally, then focus on lending to bring out nuance and character in the wine. The fruit is harvested by hand, and is destemmed with minimal crushing to avoid harsh skin and stem tannins and to preserve the fruit intensity. The must is fermented in small lots, with regular gentle pumpovers. The wine is pressed at dryness and racked to a combination of new and used French and French-coopered American oak barrels for twelve months of aging. The 1998 Mendocino Uplands Zinfandel was bottled in July 1999.

Related information:
Varietal Composition: 85% zinfandel, 10% petite syrah, 5% carignane
Alcohol: 14.3%
Total Acidity: 6.3g/L
pH: 3.75
Release Date: February 1st, 2000

Production: 2500 cases sold at winery and distributed.
Retail: $17.00

1998 Redhead Vineyard Zinfandel Redwood Valley

Yes, There is a Redhead
...for whom this vineyard is named. But she's not the only reason. Acquired by Fife Vineyards in August 1996, this vineyard is on flaming-red soil on benchland known as the Ricetti Bench, an area known for over a century as one of the prime places in Mendocino County to grow Petite Syrah and Zinfandel. At 1200 feet elevation, the site overlooks Lake Mendocino to the south, and to the north, looks down a gentle slope into the beautiful Redwood Valley. The property includes a small winery surrounded by a dry-farmed, certified-organic vineyard planted exclusively to reds, head-trained Zinfandel and Petite Syrah, tow of the county's most successful varieties.

What To Expect From This Wine
An explosion of flavors—intense berry, black plum, spices and white pepper. With concentration and power typical of hillside fruit, this wine's rich, mouthfilling texture lasts and lasts. Just when you think you know the wine, a bite of food brings out a whole new blast of flavor. Given its balance and intensity, we think it merits cellaring for up to ten years—if you can resist this Redhead's sassy, upfront charm.

Low Yield + Hang Time = Fruit Power
The 1998 vintage was "shy-bearing." It began with a fuzzy transition between the "winter that wasn't" and the "wanna-be spring". Spotty rain during the bloom period interfered with berry set and resulted in lower crop yields for many vineyards. 1998 gave us very dark Zin fruit and the long "hang time" between verasion (mid-August) and harvest (mid-October) in what was a fairly cool year allowed for development of supple tannins and rich "red fruit" flavors. This contrasts with the big tannins and "black fruit flavors" of 1997. To sum up: deep-rooted, old vine power was focused on a small crop in a generally cool year over a generous maturation period.

The Gentle Art of Winemaking
Our winemaking philosophy is simple: obtain grapes with personality, handle them minimally, then focus on lending to bring out nuance and character in the wine. The fruit is harvested by hand, and is destemmed

with minimal crushing to avoid harsh skin and stem tannins and to preserve the fruit intensity. The must is fermented in small lots, with regular gentle pumpovers. The wine is pressed at dryness and racked to a combination of new and used French and French-coopered American oak barrels for twelve months of aging. The 1998 Redhead Vineyard Zinfandel was bottled in February 2000 with. 862 cases produced.

Related information:
Varietal Composition: 89% Estate Zinfandel, 11% Estate Petite Sirah
Alcohol: 14.2%
Release Date: April 1st, 2000
Retail: $24.00

Fife Vineyards also produces Petite Sirah, Barbera, Sangiovese and Charbono.

Folie à Deux

Folie à Deux
3070 N. St. Helena Hwy.
St. Helena, CA 94574
Phone: 800-473-4454 Fax: 707-963-9223 Email: fantasy@folle-a-deux.com

Folie à Deux literally means "a shared madness of two." It is the psychiatric term for a condition in which two individuals share the same delusional ideals or fantasies about the real world. When the original owners, both psychiatric professionals, told their friends they were fulfilling a mutual dream by starting a winery, the general response was "you're both crazy." Fellow psychiatrists jokingly suggested the couple were showing the classic symptoms of Folie à Deux. Appreciative of the lighthearted nature of such responses, they couldn't imagine a more appropriate name for their new winery.

After Folie à Deux had been chosen for the name, a logo was needed. A number of Rorschach test cards were shown to local graphic artist Susann Ortega. As young girls, Susann and her twin sister had often danced together, sharing the dream of being professional dancers. She created a drawing of twin dancers in the fashion of a Rorschach "inkblot." It seemed like a perfect logo for the winery.

In 1981, Folie à Deux began producing wine from its 12 acres and in early 1985, the winery was sold to a small group of wine enthusiasts, including renowned winemaker, Dr. Richard Peterson, who now serves as Chairman. To further enhance, solidify and broaden Folie à Deux's commitment to excellence, winemaker, Scott Harvey was persuaded to join the team.

The Tasting Room is in a century-old farmhouse and is open 7 days a week from 10:00am to 5:00pm and 5:30 during daylight-savings time. Picnic grounds. Working wine cave available for private tastings and parties.

Chairman: Dr. Richard Peterson   **Winemaker**: Scott Harvey

1998 Bowman Vineyard Zinfandel, Amador County

Appellation: Amador County

Vineyard Designation: Bowman Vineyard: Chuck and Dick Bowman planted this vineyard in 1974. It sits atop the north side of the highest mountain in the center of the Shenandoah Valley of Amador County where the soil is a decomposed granite called Sierra Series Sandy Loam. The vineyard is dry farmed using the old style head-pruning method, also known as Goblet pruning.
Composition: 100% Zinfandel
Maturation: The wine spent 18 months in French and American Oak.

Related information:
Alcohol: 13.5%   **Residual sugar**: Dry
Brix at Harvest: 24   **Harvest Date**: October 20, 1998
Bottling Date: March 24th, 00   **Release Date**: April 1, 2000
Production: 2073 cases sold at winery and distributed.
Retail: $26.00

W i n e m a k e r ' s   N o t e s : The "Old Vine" complexities are apparent in this wine as are the varietal characteristics of a rich, full-bodied Zinfandel and the flavors specific to Amador County. The Bowman Vineyard traditionally produces a bright, mouthfilling and flavorful wine. Rich, dark cherry aromas and hints of anise characterize this vintage as rich fruit flavors lead to a pleasant, mint finish. The wine will make you sit up and take notice.

1998 D'Agostini Vineyard Zinfandel

Our winemaker has made Zinfandel from this 78 year-old, non irrigated vineyard for 20 years. The gnarled vines of this famous D"Agostini Brother Vineyard produce a full-bodied wine with black raspberry flavors combined with inviting aromas of cedar, plums and white pepper. This region is known for exceptional pre-prohibition era Old Vine Zinfandel vineyards. Wines from this vineyard are classic, showing a rich, full flavored complexity exemplifying the best of Zinfandel.
To be released in the summer of 2000

1998 Eschen Vineyard Zinfandel, Fiddletown "Old Vine"

The 85 year-old Eschen Vineyard is not irrigated forcing the roots to continually push deeper into the soil and extract the characteristics specific to Amador County and the Fiddletown Appellation. This particular vineyard is very low yielding at less than 1 ½ tons per acre.

Folie à Deux, Napa   100

To be released in the summer of 2000

Cellar Notes:
1995 Amador County "Old Vine": Viscous and full-bodied with black raspberry flavors combined with tantalizing aromas of cedar and plum.
1996 Amador County "Old Vine": From the D'Agostini Brother vineyard, this wine is full-bodied with black raspberry flavors combined with aromas of cedar, plums and white pepper.
1996 Eschen Vineyard, "Old Vine": Blackberry and current flavors with just a hint of anise.
1997 Amador County: The nose has blackberry, cherry and lush, jammy aromas. This full-bodied wine has a soft, yet firm tannin structure which leads to a spicy, peppery finish.
1997 Eschen Vineyard, "Old Vine": Full-bodied wine with a ruby gem color, black pepper, licorice, berry aromas and apple cinnamon flavors.
1997 D'Agostini Vineyard,: Black raspberry flavors, aromas of cedar, plum and white pepper
1997 Bowman Vineyard: Sweet cherry tobacco aromas lead to ripe fruit flavors and the long, lingering finish.
1997 Harvery-Binz Vineyard "Old Vine": Raspberry, black cherry, aromas of cedar and plum

Folie à Deux also produces Sangiovese, Cabernet Sauvignon, Chardonnay, Barbera, Syrah, Chenin Blanc, Menage à Trios, and Sparkling Wine.

Folie à Deux, Napa

Forchini Vineyards & Winery

Forchini Vineyards & Winery
5141 Dry Creek Road
Healdsburg, CA 95448
Phone: 707-431-8886 Fax: 707-431-8881
e-mail: wine@forchini.com   website: www.fochini.com

Forchini Vineyards has been growing premium varietal winegrapes in Sonoma County since 1971 from two historic vineyards located in the Dry Creek and Russian River Appellations. We are family owned and operated and produce only Estate Grown and Bottled wines. The Winery was built in 1996 at Dry Creek for 3000 case annual production.

Winery sales, tours and tasting by appointment. Direct sales to on/off premise accounts within California. On-line sales to customers in reciprocal states. Limited distribution outside California

Proprietors: Jim and Anita Forchini   **Winemaker**: Jim Forchini
Vineyard Manager: Andrew Forchini

Winemaking Style:

Estate fruit is hand picked and quickly destemmed. Primary fermentation is done in open top stainless fermenters using cultured yeast, and cap management is by punch-down. Closed stainless tanks are used for primary completion, cultured ML bacteria inoculation and settling prior to racking to oak. Aged in a combination of new American and used French oak for 12-16 months. Emphasis is on minimal processing, sanitation, and gentle handling of the fruit to avoid extracted bitterness. Polished filter prior to bottling.

1998 "Papa Nonno" Zinfandel Old Vine Clone

Appellation: Dry Creek Valley, Estate Bottled
A highly extracted wine with dark garnet color, aroma, and rich fruit flavors of berry, cherry and plum. Subtle oak extraction does not mask the perfume of fruit and bouquet. A blend of 85% Zinfandel co-fermented with 15% Carignane. Well balanced suppleness and mouthfeel without the hot after finish normally associated with high alcohol.
Related information:

Alcohol: 15.5% **Residual sugar**: Dry
TA :.570 gms/100ml **PH**: 3.52
Release Date: April 2000
Production: 960 cases
Retail: $20.00

Forchini Vineyards:
Russian River Terrace, 24 acres, acquired 1971 planted to Chardonnay, Pinot Noir, Zinfandel.
Dry Creek Bench, 67 acres, acquired 1976 planted to Cabernet Sauvignon and Zinfandel.

Frog's Leap

Frog's Leap
Post Office Box 189
8815 Conn Creek Road
Rutherford, CA 94573
Phone: 707-963-4704 Fax: 707-963-0242

Frog's Leap was founded in 1981 on a spot along Mill Creek known as the Frog Farm. An old ledger revealed that around the turn of the century frogs were raised there and sold for 33 cents a dozen, destined no doubt, for the table of Victorian San Francisco gourmets.

Now in its twentieth year of production, Frog's Leap is at home amongst 130 acres of vineyards in Rutherford at the historic Red Barn. This grand and welcoming building was built in 1884 as the Adamson Winery and renovated in 1994 as Frog's Leap's permanent home. Winemakers John Williams and Paula Siroky hand-craft an annual production of almost 50,000 cases comprised of Cabernet Sauvignon, Chardonnay, Leapfrögmilch, Merlot, Sauvignon Blanc, and Zinfandel.

Using the best of Napa Valley's organically grown grapes and the most traditional winemaking techniques, Frog's Leap strives to produce wines that deeply reflect the soils and climate from which they emanate. Frog's Leap produces some of Napa Valley's finest wines and, undoubtedly, has one of the wine world's best mottos: "Time's fun when you're having flies."

Visitors welcome Monday through Saturday 10am- 4pm. Closed Sundays, Holidays and for scheduled Special Events. Tours include wine tasting and are scheduled by reservation.

Proprietors: The Williams Family
Winemakers: John Williams and Paula Siroky

1998 Napa Valley Zinfandel

Appellation: Napa Valley
Vineyards:
St. Helena: Vineyards are located on the benchland as well as the valley floor. Organically farmed, the soil is deep to moderately deep loam with up to a 15% slope. Fertility ranges from low to vigorous with vines averaging 15-25 years old.

Rutherford: These organically farmed vineyards are located on the northern Rutherford Bench. Organically farmed the soils are moderate to vigorous fertility with deep, well-drained gravely loam at 5% to 10% slope. The vines are on average 15 to 25 years old.

Composition: 93% Zinfandel 4% Petite Sirah 3% Napa Gamay
Vinification and Maturation: Native yeast fermented, skin contact for one week to ten days. Aged for 14 months in French and American oak. Natural malolactic, unfined, unfiltered.

Related information:
Alcohol: 13.9%	**Residual sugar**: NA
Brix at Harvest: 23.7	**pH and TA**: 3.46pH 6.2 gr/1 total acidity

Harvest Date: Late September through mid-October 1998
Bottling Date: Late January 1999
Release Date: June 1, 2000
Production: 7,900 cases
Retail: $19.00

Vintage Notes:

Some regard 1998 as the heart-stopping Vintage from Hell - with a small crop that refused to ripen until late in the Fall and then only under threat of early Winter rains. 1998 provided Frog's Leap's grapes the opportunity to mature and ripen at their own pace, achieving natural complexity without the interference of high alcohol degree. 1998 gave an advantage to wines that prefer finesse to power, balance to strength and natural charm to big impressions.

Winemaker's Notes:

It's a gift from Mother Nature when a vintage arrives beautifully. This is one of those special wines that followed a natural rhythm from the moment the fruit came in, leaving little for a winemaker to do than step out of the way. The resultant wine is a beauty: great color, bright Zinfandel fruit - full of raspberries and spice - with a silky smooth, balanced flavors.

Frog's Leap also produces Sauvignon Blanc, Chardonnay, Cabernet Sauvignon and Merlot.

Gabrielli Winery

Gabrielli Winery
10950 West Road
Redwood Valley, CA 95470
Phone: 707-485-1221 Fax: 707-485-1225
www.gabrielliwinery.com

Gabrielli Winery is located at the headwaters of the Russian River in Redwood Valley, California's northern most grape growing region. Mendocino County's first vineyards were planted here in the 1850's. Situated at the frontier of the coastal rainforest, Redwood Valley receives almost twice as much rainfall as Sonoma and Napa Counties to the south. Over eons, these rains have depleted our soils of essential nutrients forcing our grapevines to struggle to produce tiny, currant-sized berries with highly concentrated flavors.

The winery was founded in 1989 by Sam Gabrielli, Bernadet Yamada-Gabrielli and Tom Yamada. Today, they, along with Joe Spliethof, Frank Pickrell, and Loretta Byrne, all contribute their individual talents to growing the wine and manning the winery and vineyard. Our estate vineyard is planted to Sangiovese and Syrah, the great varieties of Tuscany and the Rhône. Our wines are unfiltered, coaxing every drop of flavor from the grape. At Gabrielli, intensity and elegance are balanced to create wines of unparalleled quality.

Tasting Room and picnic area open Monday through Friday 10:00 am to 5:00 pm, weekends by appointment. Mail order. Wine Club. Tours by appointment.

Winemaker: Sam Gabrielli

Winemaking Philosophy: California's adopted variety, basking on hillsides, watered by the clouds. Gnarled vines, older than our president; stewarded by the Villanovas, Luvisis, and the Goforths. We concentrated our Zin's flavors by aging in small French barrels. A friendly wine loaded with raspberry and chocolate flavors. We are the first winery to produce a California wine (Nativo Zinfandel) made not only from grapes grown in Mendocino County, but also aged in barrels made from Mendocino-grown oak.

1997 Reserve Zinfandel Mendocino

Appellation: Located 120 miles north of San Francisco, Mendocino County is a microcosm of California's best grape growing regions. The

Mendocino wine region is bounded by California's Coastal Mountain Range, the Pacific Ocean, and the great northern redwood forests. **Vineyard**: Our Reserve Zinfandel is a barrel selection which emphasizes concentration, depth of fruit and integrated oak. Three dry farmed vineyards make up this wine. The Chiarito Vineyard planted during the Reagan era is comprised of four old clones from the Mendocino ridge tops each lending their own complexity. The Goforth Vineyard planted during the Nixon era is made up of one third Petite Sirah and contributes flavors of blueberries and black pepper, and the Luvisi Vineyard which was planted during the Wilson era (lest you think we make only Republican wine) gives the wine velvety raspberry notes.

Composition: 10% Petite Sirah and 90% Zinfandel

Vinification and Maturation: The 1997 growing season was long, cool, and dry. The Zinfandel was crushed, pumped over and pressed at dryness. Wild type and cultured yeast were used. The wines were pressed and immediately transferred to a combination of French, local and American oak. The wine was not filtered or fined prior to bottling in March of 1999.

Related information:
Alcohol: 14.5%	**Residual sugar**: 0.1%
TA: 7.4 gm/l	**pH**: 3.62
Release date: 3/99	
Production: 500 cases	
Retail: $25.00	

W i n e m a k e r ' s   N o t e s : Big aromas of black and red berries followed by toffee with integrated tannins and a long finish.

1997 Luvisi Vineyard, Napa Valley

Vineyard: The Luvisi family's dry farmed Zinfandel vineyard was planted around 1914 alongside the Silverado Trail just south of Calistoga. At the time this was still a wagon trail. This vineyard of old gnarled vines, three feet tall and one foot in diameter, is interplanted with figs, olives and apricots in the traditional Italian cultura promiscua way, yields small amounts of intense blue black fruit.

Composition: 100% Zinfandel.

Vinification and Maturation: The vineyard was harvested in the middle of September at 25 degrees Brix. The yield this year was 2.8 tons per acre. Open top fermentation with wild, Italian and French yeast, punched down for two weeks and pressed into a mixture of French and Mendocino oak barrels. The wine was aged in barreled for eighteen months, fined with milk protein, and was not filtered.

Related information:
Alcohol: 14.5%   **Residual sugar**: 0.2%
pH: 3.55   **Harvest Date**: Spring 1999
TA: 6.1 gm/l
Production: 1000 cases
Retail: $20.00

W i n e m a k e r ' s   N o t e s : Intense aromas of blackberry, raspberry, cracked pepper and creamy wood. Deep dusty tannins, a solid, big wine with a lingering finish.

1998 Goforth Vineyard, Redwood Valley

Vineyard: Redwood Valley in Mendocino County is a rolling, rugged, and red-soiled region. Slightly cooler and later ripening than surrounding alley, Redwood Valley has a varied terrain that creates a wide range of micro-climates in itself. The Zinfandel was planted by Ted Goforth in 1972 on Pinnobie loam in southwest Redwood Valley. The vineyard is on a hill facing northeast. It is non-irrigated, head trained and cross cultivated. It is the last Zinfandel Vineyard to be picked every year.
Composition: 15% Petite Sirah and 85% Zinfandel.
Vinification and Maturation: The zinfandel was harvested October 14, 1998. This was a very cool, late growing season. The grape acidity stayed high which meant we needed riper fruit for balance. Luckily the extra hang time was in dry weather. All the fruit was lightly crushed and destemmed. Fermentation were not inoculated. The Zinfandel was kept on the skins for two weeks after dryness and the Petite Sirah was pressed at dryness. Aging of the wine was in 50% new French oak . The wine was bottled February, 2000.

Related information:
 Alcohol: 14.0%   **Residual sugar**: 0.2%
 pH: 3.56   **Bottling Date**: 6/8/97

Production: 675 cases

Retail: $20.00

W i n e m a k e r ' s   N o t e s : Dark garnet color, nose of fresh fruit and creamy wood, cracked pepper and raspberry. Solid and sizable, with a long finish.

Gabrielli Winery also produces Estate Sangiovese, Estate Syrah, Pinot Noir and a dry rosé.

Gary Farrell Wines

Gary Farrell Wines
P.O. Box 342
Forestville, CA 95436
Phone: 707-433-6616 Fax: 707-433-9060
www.garyfarrell.com

While a political science student at Sonoma State University, Gary was intrigued by wine and got a job as a cellar worker at Davis Bynum Winery. Where better to learn the process than among the tanks, equipment and barrels of a cellar. But the art of winemaking comes from somewhere else. Maybe it is the beauty of the vineyards that inspire this expression of passion but like a storyteller, winemakers communicate a sense of place and season in the wines they craft. Letting the vineyard reveal its personality rather than the winemaker's is what makes Gary's wines so impressively honest.

Many of Gary Farrell Wines are available only through our newsletter or website. To be placed on the mailing list, please call or write at the above address or phone number, or visit the website. As of this publication, we have just finished building our new winery on Westside Road in Healdsburg, California. Tours and tasting may be available by appointment at this facility, but such appointments will be restricted to certain dates and times.

Owner and Winemaker: Gary Farrell

Winemaking Philosophy:
It is all in the vineyard. Finding the right source and working with those who grow the grapes, who are the caregivers of that vineyard, is perhaps my most important task as a winemaker. Handling the grapes gently, from harvest to bottling, respecting the "personality" of the grape so that it shows through in the wine is simply what it is all about. My focus for many years has been Pinot Noir and it has been only in the last few years that my interest in Zinfandel flourished.

Vinification and Maturation:
The wines are made with 100% Zinfandel. The grapes are hand picked and hand sorted and then cold soaked prior to fermentation. Using small open top fermentors that hold 2 tons, fermentation takes approximately nine days. A slow fermentation is preferred using the natural yeast that comes in from the vineyard although Assmannshausen yeast is sometimes supplemented to insure a complete

fermentation. Punch down cap management is used for a more gentle handling of the young wine. Malolactic fermentation occurs in barrels with some lees present. Only one racking is performed. I believe in handling the grapes and the wine gently with as little agitation as possible throughout the wine making process. The wine is then aged 14 months in 60 gallon French oak barrels, 35% to 45% new. This approach to vinification and maturation is consistent with all my Zinfandel wines.

Current Releases

1996 Sonoma County, Old Vine Selection

This is a visually breathtaking wine, with its youthful deep purple hue. Its intense, deeply filled aromas of raspberries, blackberries and creamy oak lead to attention-getting flavors that are insatiably long and wonderfully proportioned. Our source for this Zinfandel include some of the most recognized Zinfandel vineyards in Sonoma County...the Collins or Limerick Lane Vineyard and the Bradford Mountain Wine Creek Vineyard. A third new source in this blend is the Ash Creek Vineyard, which is the highest elevation Zinfandel vineyard in the country at 2200 feet above sea level.

    100% Sonoma County
    40% Bradford Mountain-Wine Creek Vineyard
    40% Pine Mountain-Ash Creek Vineyard
    20% Collins or Limerick Lane Vineyard

Related information:
Alcohol: 13.8%   **Brix at Harvest**: 23.8
Harvest Date: Late September
Bottling Date: 11-97   **Release Date**: 05-98
Production: 1350 cases   **Price**: 22.50

1997 Sonoma County, Old Vine Selection

    100% Sonoma County
    19% Bradford Mountain-Wine Creek Vineyard
    53% Pine Mountain-Ash Creek Vineyard
    28% Collins or Limerick Lane Vineyard

Related information:
Alcohol: 13.7%   **Brix at Harvest**: 23.7
Harvest Date: September 12th and 25th
Bottling Date: 11-98   **Release Date**: 05-99
Production: 950 cases   **Price**: 24.00

1997 Dry Creek Valley, Maple Vineyard

We feel that the Maple Vineyard is perhaps the most important and exciting new Zinfandel source to come our way in quite some time. For many years the Maple Vineyard was one of the three primary sources for Ridge Winery's Lytton Springs Zinfandel. Our new association with Tom and Tina Maple will give us a unique opportunity to select the finest Zinfandel blocks from their vineyard each year for our production. This is an incredibly important relationship for our winery, given our long-term commitment to this varietal.
100% Sonoma County/Dry Creek Valley
100% Maple Vineyard

Related information:
Alcohol: 13.8%   **Brix at Harvest**: 23.9
Harvest Date: September 8th
Bottling Date: 11-98   **Release Date**: 10-99
Production: 300 cases   **Price**: $30.00

1997 Dry Creek Valley, Bradford Mt.

Another fabulous property and new Zinfandel source is the Bradford Mountain vineyard. The red volcanic soils of this location provide for powerful wines of immense concentration and structure. This wine is no exception with its bold and noble lines. It is deep and rich, with nuances of fresh black fruits, pepper and spice.
100% Sonoma County/ Dry Creek Valley
100% Bradford Mountain
100% Hambrecht Vineyard

Related information:
Alcohol: 13.8%   **Brix at Harvest**: 23.8
Harvest Date: September 12th
Bottling Date: 11-98   **Release Date**: 10-99
Production: 320 cases   **Price**: 28.00

Upcoming Releases for 2000

1998 was an outstanding vintage for Zinfandel in Sonoma County's Dry Creek Valley. The vines set an unusually small crop due to poor weather during bloom. The growing season was very cool, which led to an extremely late harvest. Fortunately, there was no rain during the period, and the grapes were harvested in exceptional condition. The delayed maturation and limited yields provided for maximum flavor development, profound intensity and superb structure in the wines produced from this vintage.

1998 Sonoma County, Dry Creek Valley
Vineyard Designation: None
Brix at Harvest: 23.9
Bottling Date: 11-99
Release Date: 06-00
Production: 1425 cases

1998 Ricci Vineyard, Russian River Valley
Vineyard Designation: Ricci Vineyard
Brix at Harvest: 23.5
Bottling Date: 11-99
Release Date: Undetermined
Production: 500 cases

1998 Bradford Mountain, Dry Creek Valley
Vineyard Designation: Bradford Mountain
Brix at Harvest: 23.9
Bottling Date: 11-99
Release Date: Undetermined
Production: 300 cases

1998 Maple Vineyard, Dry Creek Valley
Vineyard Designation: Maple Vineyard
Brix at Harvest: 23.8
Bottling Date: 11-99
Release Date: Undetermined
Production: 650 cases

Gary Farrell also produces Pinot Noir, Chardonnay, Merlot and Cabernet Sauvignon.

Geyser Peak Winery

Geyser Peak Winery
2281 Chianti Road
Geyservillle, CA. 95441
Phone: 707-857-9463 or 800-945-4447 Fax: 707-857-3545

Established in 1880 by pioneering winemaker Augustus Quitzow, Geyser Peak Winery is making some of the finest and most awarded wines in the world. Located in the heart of the Alexander Valley in Sonoma County, Geyser Peak Winery boasts a state of the art winery, including rotary fermenters and a winemaking team dedicated to producing outstanding wines.
Tasting Room and picnic facilities open daily from 10:00 to 5:00. Mail order. Wine Club.
Owner: Jim Beam Brands Co.
Winemaker: Daryl Groom and Mick Schroeter

Winemaking Philosophy:
Geyser Peak's 1998 Zinfandel was produced using fruit primarily from the Alexander Valley of which approximately 75% was from old vines. A small portion from the Cucamonga Valley was added to give some intense richness and concentration. These vines are over 120 years old. Our focus for Zinfandel is to have elegance and harmony to express all of the varietal characteristics that are typical for Zinfandel.

1998 Geyser Peak Zinfandel

Appellation: Sonoma County
Vineyard: We have a core of four growers that we work with every year to help produce the best quality grapes for our Zinfandel. Three of these are considered to have old vines ranging from 60 to 120 years old.
Composition: 98% Zinfandel and 2% Shiraz
Vinification and Maturation: The grapes were fermented in individual lots in rotary fermenters with only five days of skin contact achieving rich extraction and color without excessive tannin. Once the wine was pressed and finished malolactic fermentation the best lots were blended together and aged in a combination of French and American oak barrels for 14 months.

Related information:
Alcohol: 14.5 %   **Residual sugar**: Dry
Brix at Harvest: 24.0-25.0
Bottling Date: April 2000   **Release Date**: June 2000
Production: 1,920 cases sold at winery and distributed.
Retail: $17.00

Winemaker's Notes:

The 1998 Geyser Peak Zinfandel is a wine that exhibits the best flavors that Zinfandel can offer. Although there was a relatively cool growing season, we were able to achieve full ripeness in all of our Zinfandel vineyards providing dense, rich, and concentrated wines. The wine shows a great jammy character with spicy and cedar notes. A slight oak, vanillin character rounds out the aromas and the palate. This wine is wonderful in its youth, but has a structure and complexity that will allow it to age gracefully for 3- 5 years as well.
Food pairing: Almost any pasta dish, gamy meats, and perfect for the BBQ.

Geyser Peak Winery also produces Chardonnay, Sauvignon Blanc, Gewürztraminer, Riesling, Cabernet, Merlot, Shiraz and Meritage.

Geyser Peak Winery, Sonoma County

Green and Red Vineyards

Green and Red Vineyards
3208 Chiles Valley Road
St. Helena, CA 94574
Phone: 707-965-2346

Green and Red Vineyard founded in 1977 and named for its red and iron soils veined with green serpentine, is located in the steep hills on the east side of Napa Valley. Planting was started in 1972 on ground that was originally planted as vineyard in the 1890's. Our three vineyards range in elevation from 1000 to 2000 feet with different exposures.

Tours on Saturdays from 1:00 pm to 4:00 pm by appointment only.

Owner, Winemaker, Vineyard Manager: Jay Heminway

Winemaking Philosophy:

The emphasis is on the fruit. In 1972, seven acres of Zinfandel were planted in the Chiles Mill Vineyard overlooking Chiles Canyon where remnant walls of the grist mill built by pioneer J.B. Chiles still stand. The vines on this hillside shelf produce a vigorous, black pepper and berried wine. Cropped to three tons per acre with selective leaf pulling and cluster thinning along with gentle handling in the winery enhance the focused intensity of this fruit.

Vinification and Maturation:

Hand picked and hand sorted grapes. Three to four days cold soaked prior to primary fermentation. Of the destemmed fruit, 40% is whole berry. Open top, cool fermentation with native yeast and hand punch down for ten days. Aged for twelve months in 20% American oak - 10% new and 80% in French oak.

1998 Zinfandel Chiles Mill Vineyard

Appellation: Napa Valley
Vineyard Designation: Chiles Mill Vineyard, Estate. (29 tons)

Related information:

Alcohol: 13.8%	**Residual sugar**: Dry
Brix at Harvest: 24	**Harvest Date**: October 12th -26th
Bottling Date: 12-15-99	**Production**: 1954 cases
Retail: $22.00	

W i n e m a k e r ' s   N o t e s : Raspberry, blackberry laced with cherry vanilla custard dotted with black pepper and cinnamon. Opulent now, more opulent later.

1998 Zinfandel Chiles Valley Vineyards

Appellation: Napa Valley
Vineyard: The grapes used to make this wine come primarily from vineyards in neighboring Chiles Valley at 800 to 1,000 feet elevation. 14 tons picked.

Related information:
Alcohol: 13.9%	**Residual sugar**: Dry
Brix at Harvest: 24	**Harvest Date**: October 18th - 28th
Bottling Date: 12-14-99	
Production: 938 cases	
Retail: $20.00	

W i n e m a k e r ' s   N o t e s : Red raspberry, black cherry pie with dark crust and spices with black pepper and minty clove.

Green and Red Vineyards also produces a Napa Valley Chardonnay from Catacula Vineyards and a Napa Valley Gamay.

Grgich Hills Cellar

Grgich Hills Cellar
P.O. Box 450
1829 St. Helena Hwy.
Rutherford, CA. 94573
Phone: 707-963-2784 Fax: 707-963-8725

Grgich Hills Cellar, located on Highway 29 just north of Rutherford, was formed in 1977 through the efforts of Miljenko Grgich and Austin Hills of the Hills Bros. coffee family. Back in 1977 Grgich had the skill and expertise to make great wines, and Hills, owner of established vineyards, had a background in business and finance. Excellence is an uncompromising philosophy at Grgich Hills Cellar. Each wine is created as a special child, receiving devoted attention to every stage of its development to achieve its own unique character. Never satisfied with yesterday's best, each vintage arrives as a new challenge to create the best wine of the future.

Tasting Room open daily from 9:30 am to 4:30 pm.
Proprietors: Miljenko Grgich and Austin Hills
Winemaker: Miljenko "Mike" Grgich

Winemaking Philosophy:
I was taught to treat grapes and wines as if they were living things - like children. So, it is no wonder that I often refer to a wine as having five lives or stages in development.

The first life begins in the fertile soil of the vineyard where the varietal character of the fruit develops. Next, the actual birth of the wine occurs in the cellar during fermentation. Third, the newborn is cradled in oak barrels in order to grow and develop. The fourth stage occurs during bottle aging when the wine has matured sufficiently or come of age. During this period the flavors of the wine and oak begin to marry to create the bouquet, the highly prized breath of fine wine.

Finally, finished with a silken texture, refined complexity and sophisticated individuality, the wine is ready for enjoyment. It is our belief at Grgich Hills that fine wines can only be created by recognizing the unique requirements of each of these five distinct lives.

1997 Zinfandel, Sonoma County

Vineyard: A portion of the Zinfandel is from old vines with the balance selected from middle-aged vines.

Composition: 100% Zinfandel

Vinification: Hand picked. Cold fermentation with gentle pump over, drained and pressed at .02% residual sugar to maintain freshness and fruitiness. Fifty percent partial malolactic fermentation in three and four year old barrels to preserve the freshness of the fruit.

Maturation: Thirteen to fourteen months in 75% older American oak and 25% newer French oak. One year or more of bottle age.

Related information:

 Alcohol: 13.7%   **Residual sugar**: 0.02%

Brix at Harvest: 23.5 - 24.5

Harvest Date: September 1997

Bottling Date: 9/99   **Release Date**: April 2000

Production: 8700 cases sold at winery and distributed.

Retail: $23.00

Winemaker's Notes:

With Zinfandel being as popular as it is these days, there just doesn't ever seem to be quite enough to go around, so we thought we'd do our part for the cause by offering you this luscious hedonistic wine.

Distinctively true to its type, our 1997 Zinfandel has a highly aromatic, spicy, briary nose that entices you to sample it without discussion or delay. The big, mouth-filling flavors that follow will tickle your taste buds and make you think of sitting at the kitchen table hovering over a slice of warm blackberry pie with melting vanilla ice cream. The marvelous aroma and explosive taste of this wine are followed by a long, lingering aftertaste that will leave you lusting for more.

Grgich Hills Cellars also produce a Chardonnay, Fume Blanc, Cabernet Sauvignon and Violetta (a blend of Riesling and Chardonnay, Violetta is a late harvest dessert wine).

Gundlach Bundschu Winery

Gundlach Bundschu Winery
P.O. Box 1 Vineburg, CA 95487
Tasting Room: 2000 Denmark Street
Sonoma, CA. 95476
Phone: 707-938-5277 Fax: 707-938-9460
wino@gunbun.com

Nestled in the southern-most corner of the Sonoma Valley and founded in 1858 by Jacob Gundlach and son-in-law, Charles Bundschu, the 400 acre Rhinefarm Vineyards is home to one of California's oldest and most respected wineries. Now owned and operated by fifth and sixth generation Bundschu's, Gundlach Bundschu winery produces 55,000 cases of premium varietal wines annually, with Zinfandel becoming a part of the line-up in the 1970's.

Tasting Room and picnic grounds open daily from 11:00 to 4:30. Mail order. Wine Club. Seasonal Tours. Summer Shakespeare Festival.

Proprietor: Jim Bundschu   **Winemaker**: Linda Trotta

Winemaking Philosophy:
Both of Gundlach Bundschu's vineyard designated Zinfandels are crafted in a style that highlights the unique varietal characteristics of each ranch. The Estate Rhinefarm Vineyards fruit is characterized by deep blackberry fruit accented with a distinctive black pepper quality. From the Morse Vineyard, located in Glen Ellen, comes fruit with a ripe raspberry fruit character.

Vintage Notes:
1998 provided a relatively cool growing season and a late start to the harvest on Rhinefarm. Because of the mild temperatures and merciful absence of rain during the harvest period, the fruit enjoyed a long time on the vines to mature and ripen. The overall effect on the wines is a delicate balance on the palate. The longer ripening time allowed the grape tannins to mature and integrate with the fruit, thus producing wines of wonderful balance and silky texture.

1998 Zinfandel Estate Rhinefarm Vineyards

Appellation: Sonoma Valley **Vineyard Designation**: Estate Rhinefarm Vineyard. Located on our estate, this is a south-west facing hillside at the base of the Mayacamas Mountain Range.

Composition: 100% Zinfandel

Vinification and Maturation: Hand picked and fermented in close-top fermenters for 10 to 15 days, with two pump overs per day. The wine ages for 12 to 14 months in American oak barrels and is bottled filtered and unfined.

Related information:
Alcohol: 14.9%	**Residual sugar**: Dry
Brix at Harvest: 24.9	**Harvest Date**: 10/7/98
Bottling Date: 01/00	**Release Date**: May 2000
Production: 2000 cases.	

W i n e m a k e r ' s   N o t e s :
Fresh ground pepper and raspberry fruit accented by toasted oak fill the nose of this big Zin. Flavors of brambly fruit, raspberry jam and spicy oak integrate with the tannins to create a rich middle and lively finish.

1998 Zinfandel Morse Vineyards

Appellation: Sonoma Valley

Vineyard Designation: Morse Vineyard. Our Sonoma Valley Zinfandel is produced from grapes grown on the Morse Vineyard in Glen Ellen, on the eastern edge of the Sonoma Valley.

Composition: 100% Zinfandel

Vinification and Maturation: Hand picked and fermented in close-top fermenters for 10 to 15 days, with two pump overs per day. The wine ages for 12 to 14 months in American oak barrels and is bottled filtered and unfined.

Related information:
Alcohol: 13.8	**Residual sugar**: Dry
Brix at Harvest: 24.1	**Harvest Date**: 10/9/98
Bottling Date: 2/00	**Release Date**: May 2000
Production: 1,200 cases sold at winery and distributed.	

W i n e m a k e r ' s   N o t e s :
This wine is full and rich while expressing the fruit in an elegant style. Opening with ripe red cherry and raspberry fruit on a backdrop of vanilla oak, the wine

glides over the palate with gentle tannins and finishes with cherry fruit and spicy oak undertones.

Gundlach Bundschu also produces Gewürztraminer, Riesling, Chardonnay, Gamay Beaujolais, Cabernet Franc, Merlot, Cabernet Sauvignon and Pinot Noir. Red Bearitage and Polar Bearitage, a red and white table wine.

GustavoThrace

GustavoThrace
880 Vallejo Street
Napa, CA 94559
Phone: 707-257-6796 Fax: 707-257-7001
Email: gustavot@napanet.net
Website: www.gustavothrace.com

GustavoThrace Winery is a small producer of fine wines. Started in 1996, the winery, located in downtown Napa, produces Zinfandel, Chardonnay and Cabernet Sauvignon.

Partner and Winemaker, Gustavo Brambila, was born in San Clemente, Jalisco, Mexico, moving to the Napa Valley as a child. After graduating from St. Helena High School he attended The University of California at Davis where he earned his degree in Fermentation Science. He began his career in 1976 learning to make wine first at Chateau Montelena and then at Grgich Hills.

Gustavo and Thrace met in 1996, establishing the winery with the first harvest of Zinfandel that fall. In 1997 Chardonnay and Cabernet were added. Total production for the winery is less than one thousand cases per year.

Visitor are welcome to stop by the winery for tasting and retail sales by appointment.

Partners: Gustavo Brambila and Thrace Bromberger

Winemaking Philosophy:

The statue on the label is known as the "Winged Victory" or "Nike of Samothrace." She is from the second century A.D. and was found on the Greek island of Samothrace (the city for which Thrace is named). In Greek mythology, Nike was the Goddess of Victory. We have chosen her as our symbol to reflect the personal dreams and victories that we achieve in bringing these wines to you. Our hope is that you will allow our wines to become a part of celebrating your own personal victories - be they very special events or simply getting through the day and enjoying life.

1997 Zinfandel, Napa Valley

Vineyard: Brown Estate Vineyards, Chiles Valley
Composition: 100% Zinfandel

Vinification and Maturation: The grapes were harvested in October of 1997. Following a long, temperate ripening period there was a temperature spike that elevated sugars throughout the valley. The fruit was crushed into half-ton bins, allowing for an 80% whole berry fermentation (open-top) to ensure the extraction of the aromas and flavors we tried to capture. Gustavo inoculated the fruit at 50 degrees but fermentation didn't begin for 2-3 days. Punch downs were done by hand three times a day through nine days of fermentation. The wine was then aged in a combination of French and American aged oak. After seven months in used barrels and extensive aeration took place to open up the wine before it was transferred to new French and American barrels for a total of seventeen months barrel aging. The result is an intense, focused Zinfandel that truly showcases the Chiles Valley where it was nurtured.

Related information:
Alcohol: 13.8%
Release Date: May 6, 2000
Cellaring: 3-7 years
Production: 350 cases produced

Winemaker's Notes:
The fruit for this Zinfandel was from the same three Chiles Valley vineyard blocks that we used to produce our 1996 Zinfandel. Gustavo has focused on the distinct profiles these individual blocks have to arrive at the final blend. One block has pronounced black pepper, one a rich spiciness and one a forward blueberry-raspberry. Together they blend into a layered intriguing nose that continues unfolding in the mouth, allowing you to appreciate the enhancement of oak aging. The result is a rich balanced finish that makes for a memorable wine.

Haywood

Haywood
P.O. Box 182
Sonoma, CA 95476
Phone: 707-252-7998 Fax: 707-252-0392

Peter Haywood began planting the Los Chamizal Vineyard near the town of Sonoma in 1976. Prior to this, it took Peter three years to clear the thickets and prepare terraces out of the rugged hills he knew would be perfect for growing low yield, intensely-flavored fruit. He crushed the first grapes for commercial production in 1980, when he founded Haywood Winery. By 1987 Peter had significantly increased production of estate wines from this prized Sonoma Valley vineyard. Haywood Winery joined the Racke USA marketing portfolio in 1991.

Haywood wines can be tasted at The Corner Store in Sonoma (707-939-0579), open daily from 11:00 to 5:30 or at the Buena Vista Historic Tasting Room 18000 Old Winery Road in Sonoma (707-938-1266 or 800-926-1266) open daily from 10:30 - 5:00. Wine Club. Mail order.

Owner: Racke USA **Winemaker**: Judy Matulich-Weitz
Consultant: Peter Haywood

Winemaking Philosophy:
As a grower and vintner, our work begins in the field. Our wines capture the essence of the unique vineyards from which they are grown.—Peter Haywood

In keeping with Peter's philosophy, my job as Winemaker is to reflect the special character of these vineyards in the wines. Fermentation is completely natural; nothing is added. I closely monitor the temperature and pump over throughout fermentation to achieve maximum color extraction. This is followed with extended maceration, leaving the juice on the skins for 20 to 35 days to soften the tannins. Each vintage is then stored in new one-year-old American oak barrels, the length of time varying from year to year, insuring the oak works well with the wine as compliment to the final product.—Judy Matulich-Weitz

1997 Haywood Estate Zinfandel Rocky Terrace

Appellation: Sonoma Valley   **Vineyard Designation**: Rocky Terrace

Composition: 100% Zinfandel

Vinification and Maturation: Hand picked and cluster selected over a four-week period. Fermented using ambient yeast, thirty day extended maceration. Free run and press wine goes directly into barrels. We rack the wine four times while aging it for twenty four months in 100% new that are personally selected from five different coopers. It is unfiltered and unfined.

Related information:

Alcohol: 14.5%	**Residual sugar**: Dry
Brix at Harvest: 23.6	**Harvest Date**: 9/11/97-10/1/97
Bottling Date: 12/20/99	**Release Date**: Projected release July, 2000

Production: 953 cases sold at both tasting room locations
 Limited distribution.
Retail: $35.00

Winemaker's Notes: This classic Zinfandel is made from fruit grown in the fractured rock terraces of Los Chamizal Vineyard on the Mayacamas Mountain, located high above the historic Sonoma Plaza. The wine is well integrated with peppermint, raspberry, and a spicy oak nose. The forward flavors of pepper and raspberry-peppermint are followed by an elegantly structured mid palate with a touch of vanilla on the finish. This creates a well-balanced round mouth feel. A limited amount of this hand-crafted wine is made each year.

1997 Haywood Estate Zinfandel Los Chamizal

Vineyard: Los Chamizal, Sonoma Valley

Our winemaking began in the vineyards as we cluster select only the fruit which showed flavor, maturity, and sugar ripeness. The individual climates of the many terraces at Los Chamizal, as well as the natural Zinfandel variation in ripening, require that we selectively harvest the crop over a four week period.

Composition: 100% Zinfandel

Vinification and Maturation: Hand picked and cluster selected over a four week period. Some extended maceration. Twenty four months in 70% American, 30% French oak, minimal filtration and unfined.

Related information:

Alcohol: 14.5%	**Residual sugar**: Dry

Brix at Harvest: 23.4   **Harvest Date**: 9/9/97-10/01/97
Bottling Date: 12/16/99   **Release Date**: April, 2000
Production: 3630 cases sold at both tasting room locations.
Limited distribution.
Retail: $25.00

W i n e m a k e r ' s N o t e s : This elegant style Zinfandel begins with aromas of red cherry and berry jam, developing into blackberry with hints of leather, cedar, and cigar box. Oak and pepper notes are present in the rich raspberry fruit flavor. The final result is a velvet textured, sophisticated wine with a long, integrated finish.

1997 Haywood Estate Zinfandel Morning Sun

Vineyard: Morning Sun, Sonoma Valley
Composition: 100% Zinfandel
Vinification and Maturation: Hand picked and cluster selected over a four week period. Some extended maceration. Twenty four months in 100% new American oak, unfiltered and unfined.

Related information:
Alcohol: 14.5%   **Residual sugar**: Dry
Brix at Harvest: 23.9   **Harvest Date**: 9/11/97-10/1/97
Bottling Date: 1/19/00   **Release Date**: Projected release July, 2000
Production: Limited distribution.
Retail: $30.00

W i n e m a k e r ' s N o t e s : This rich Zinfandel begins with aromas of cedar, blackberry, and cracked white pepper, developing into hints of black currant, with an underlying smoky presence. This wine consists of rich black fruit and spiced oak, which combine with a nice acid balance. A coffee and smoky character linger in the finish.

1997 Haywood Estate Zinfandel Shorenstein

Vineyard: Shorenstein, Sonoma Valley
Composition: 100% Zinfandel

Vinification and Maturation: Hand picked and cluster selected over a four week period. Some extended maceration. Twenty four months in 100% new American oak, unfiltered and unfined.

Related information:

Alcohol: 14.5%	**Residual sugar**: Dry
Brix at Harvest: 23.6	**Harvest Date**: 9/9/97-10/1/97
Bottling Date: 1/19/00	**Release Date**: Projected release July, 2000
Production: Limited distribution.	
Retail: $30.00	

W i n e m a k e r ' s   N o t e s : The Shorenstein is a dark maroon red color. This wine has sophisticated coffee and liquor flavors. Traces of mocha, tobacco and leather hints blend nicely with the chewy tannins and long finish.

Haywood also produces Chardonnay, Cabernet Sauvignon, and Merlot. The grapes are hand selected from some of California's finest vineyards that reflect the same care and commitment to quality that goes into Peter Haywood's famed estate hillside vineyards in Sonoma Valley. Peter Haywood endorses these wines for their expression of true varietal character, bright fruit flavors and soft oak tannins.

Some older vintages of Haywood Rocky Terrace and Los Chamizal are also available at both tasting room sites.

Hendry

Hendry
3104 Redwood Road
Napa, CA 94558
Phone: 707-226-2130 Fax: 707-226-1345

Hendry is located on the bench lands between Napa's Carneros and Mt. Veeder viticulture districts. The 120 acre Hendry Vineyard is known for producing exceptional quality fruit that is sold to several highly regarded wineries. Starting with the 1992 harvest, small lots of fruit from the best vineyard blocks were held back for the new Hendry label.

No tasting room. Please call the winery to find out where you may purchase Hendry wines.

Proprietors: George Hendry, Susan Ridley, and Jeff Miller
Grape Grower and Winemaker: George Hendry

Winemaking Philosophy: Over a period of 40 years, George Hendry observed significant differences in the wines made from fruit grown in different parts of his vineyard. This is attributed to subtle differences in soil and climate, or terroir. So the vineyard is divided into small "blocks" with each block representing a different terroir. The goal of Hendry winery is to produce wines with unique personalities that reflect the terroir of the vineyard block in when they grow. Thus, the Hendry wines are block designated. At Hendry, we believe that great wines result from a delicate balance of terroir, viticulture and enology. Our mission is to continually strive to improve that balance.

1998 Zinfandel Block 28

Appellation: Napa
Vineyard: Hendry Vineyard Block 28.
Composition: 100% Zinfandel
Vinification and Maturation: Alcoholic fermentation was completed in approximately ten days in closed stainless steel fermentation tanks with pump over two times a day. After approximately four weeks maceration, the wine was pressed and placed in barrel for malolactic fermentation. Aging was twenty-one months in 100% French oak, 35% new French oak.

Related information:

Alcohol: 14.8%	**Residual sugar**: Dry
Brix at Harvest: 25.5	**Harvest Date**: 10/03/98
Bottling Date: 6/15/2000	**Release Date**: 9/15/2000
Production: 400 cases distributed.	

W i n e m a k e r ' s   N o t e s : The flavor profile is complex and concentrated. It includes black currant, blackberry, blueberry and cherry in addition to black pepper and spice. The wine has substantial body and a firm structure. The French oak is well integrated and adds a layer of complexity and refinement to the wine. This wine will age well.

Food pairing: Beef, lamb and flavorful reduction sauces. This is a big wine that will stand up to big, complex flavor combinations.

1998 Zinfandel Block 7

Appellation: Napa
Vineyard: Hendry Vineyard, Block 7.
Composition: 100% Zinfandel
Vinification and Maturation: Alcoholic fermentation was completed in approximately ten days in closed stainless steel fermentation tanks. After approximately four weeks maceration, the wine was pressed and placed in barrel for malolactic fermentation. Aging was eighteen months in 100% French oak, 33% new French oak.

Related information:

Alcohol: 14.7%	**Residual sugar**: Dry
Brix at Harvest: 25.3	**Harvest Date**: 10/06/98
Bottling Date: 3/15/2000	**Release Date**: 6/15/2000
Production: 1,600 cases distributed.	
Retail: $24.00	

W i n e m a k e r ' s   N o t e s : The densely concentrated fruit flavors include blackberry and black cherry in addition to black pepper and spice. The wine has a round mouth feel, substantial body, and the supple tannins that are characteristic of Block 7. The French oak is well integrated and adds a layer of complexity and refinement to the wine.

Food pairing: Pork, flavorful pasta dishes or grilled vegetables.
C e l l a r   N o t e s :

1995 Hendry Block 7: The fruit is holding up, but this wine will show best if you drink it before 2001.

1995 Hendry Brandlin Vineyard: This wine will continue to show well through 2003. 1995 was the last bottling from the Brandlin Vineyard.

1996 Hendry Block 7: This was a year of big fruit extraction and the wine will last through 2002.

1997 Hendry Block 7: The 1997 vintage resulted in a huge, aggressive style that will be enjoyed through 2003.

Hendry also produces a Cabernet Sauvignon, Chardonnay and Pinot Noir.

HENDRY

Hendry Block 7

Napa Valley

ZINFANDEL

1998

ALCOHOL 14.4 PERCENT BY VOLUME

Hop Kiln Winery

Hop Kiln Winery
6050 Westside Road
Healdsburg, CA 95448
Phone: 707-433-6491 Fax: 707-433-8162

A Russian River Valley Landmark with its three drying towers, Hop Kiln Winery is surrounded by sixty five acres of vineyard. Founded in 1975, Hop Kiln has established a tradition of excellence in producing rich, full bodied red wines, most notably its Primitivo Zinfandel. Annual production is 10,000 cases with an emphasis on small lots of carefully crafted wines.

Tasting Room and picnic area open daily from 10:00 am to 5:00 pm. Mail order.

Proprietor: L. Martin Griffin, Jr.   **Winemaker**: Steve Strobl, 1984

Winemaking Philosophy: Hop Kiln believes most of the winemaking is done in the vineyard with the emphasis on the fruit aroma and flavors.

1997 Zinfandel

Appellation: Sonoma County.
Composition: 100% Zinfandel
Vinification and maturation: Hand picked and fermented in both open top tanks with punch-down and closed top with pump-overs. After fermentation, racked to used French oak barrels for 8 months.

Related information:
Alcohol: 14.5%   **Residual sugar**: 0.094%
Brix at Harvest: 23.2   **Harvest Date**: 9/17/97
Bottling Date: 8/6/98   **Release Date**: 11/99
Production: 1,871 cases sold at winery and distributed.
Retail: $18.00

Winemaker's Notes:
The wine has a deep red color with very jammy/raspberry tones and briary/brambly aromas. Pair with hearty meats/stews and roasted vegetables.

1997 Primitivo Zinfandel

Appellation: Sonoma County.
Composition: 100% Zinfandel
Vinification and maturation: Hand picked and fermented in both open top tanks with punch-down and closed top with pump-overs. After fermentation, racked to used French oak barrels for 8 months.

Related information:
  Alcohol: 14.4%   **Residual sugar**: 0.057%
Brix at Harvest: 23.0   **Harvest Date**: 9/23/97
Bottling Date: 8/7/98   **Release Date**: 12/98
Production: 656 cases sold at winery and distributed.
Retail: $22.00

Winemaker's Notes:
Medium garnet color with floral, strawberry flavors and aromas. Has a soft fruity entry with hints of leather and spice.

1998 Vintage

Not yet released, our two vineyards of '98 Zinfandel will be unique for their first time designation by vineyard - Old Windmill Vineyard and Primitivo Vineyard. These wines have received special handling both in the fermentation regimen and barrel choice to optimize their own particular character of age, clone, rootstock, geography, and terroir.

1998 Old Windmill Vineyard Zinfandel

Appellation: Sonoma County.
Composition: 100% Zinfandel

Related information:
  Alcohol: 14.7%   **Residual sugar**: 0.098%
Brix at Harvest: 23.3   **Harvest Date**: 10/01/98
Bottling Date: 3/14/2000   **Release Date**: Spring 2001
Production: 1089 cases
Retail: $25.00

Winemaker's Notes:
This 1998 bottling marks the first of a designated zinfandel named for the windmill located in the middle of the vineyard. The wine is distinctive because of the older age of the vines, their hillside location and particular terroir. As we believe much of a wine is made in the vineyard, this designation program will allow you and us to follow these wines/vines from year to year. Huge brambly/raspberry with chocolate and briary/ floral perfume.

1998 Primitivo Vineyard Zinfandel

Appellation: Sonoma County.
Composition: 100% Zinfandel

Related information:
Alcohol: 14.93%	**Residual sugar**: 0.094%
Brix at Harvest: 23.9	**Harvest Date**: 10/01/98
Bottling Date: 3/15/2000	**Release Date**: Spring 2001
Production: 1089 cases	
Retail: $24.00	

Winemaker's Notes:
This 1998 bottling is the counterpart of our vineyard designated zinfandel program. Replanted on the original 1880's vineyard benchland, this block is unique because of its young and developing age, flat ground and particular terroir. A port-like wine, redolent of raspberry jam in the aroma and flavor.

Hop Kiln Winery also produces Chardonnay, Riesling, Sauvignon Blanc, Valdiguié, Marty Grifffin's Big Red, A Thousand Flowers and a Late Harvest Zinfandel.

Joseph Swan Vineyards

Joseph Swan Vineyards
2916 Laguna Road
Forestville, CA 95436
Phone: 707-573-3747 Fax: 707-575-1605

Joseph Swan Vineyards was established in 1969 by Joseph Swan, a retired airline pilot. He purchased land in the Russian River Valley in 1967, developing a ten acre vineyard devoted to Pinot Noir and Chardonnay. His first wines were Zinfandel and sold primarily to friends. But these bold Zinfandels, with their rich fruit and age worthiness drew a following beyond the circle of friends and it wasn't long before his wines became a beacon, drawing in and inspiring other young winemakers. Many of those who apprenticed with him are now highly regarded in the wine industry.

Joseph Swan believed in the importance of the vineyard. After his death in 1989, his son-in-law, Rod Berglund, became the winemaker. Rod creates his own style but believes as Joseph did: The vineyard is everything.

Tasting Room open on weekends from 11:00 to 4:30. Mail order.

President and Winemaker: Rod Berglund

Winemaking Philosophy:

It begins and ends in the vineyard. I believe in allowing the nature of the vineyard to come through by not attempting to change the character of the grapes through winemaking techniques. Being a true "terriorist", I spend a great deal of time just walking the vineyard, getting a feel for the season, and tasting the grapes to get a sense of the personality of the vineyard.

Every vineyard is different. Some sites are like a single note, and blending with other vineyards creates the chorus, the harmony required in a wine. Other vineyard sites are so distinctive, the grapes of such high quality, they stand alone and blending them with other sites would diminish their uniqueness. All of our vineyards are quite old, ranging in age from 50 to over 120 years.

1998 Lone Redwood Ranch, Russian River

Lone Redwood Ranch is one of those old fashioned vineyards in which a large number of varieties were planted. Although I have no direct experience with the others, both the pinot noir and zinfandel seem to thrive here. Although not as old as our other zinfandel vineyards, the vines are fully mature and sufficiently old that by some standards they would be considered old vines. They certainly produce wines with plenty of stuffing and character.

Compostion:100% Zinfandel

Vinification: The grapes are hand picked by my own crews. Grapes are selected and sorted in the vineyard. Open top fermenters with 25% whole clusters. Using whole clusters helps moderate fermentation. The juice is released slowly from the whole berries and helps maintain a cooler temperature in the tanks. I punched down by hand several times per day, as this is a gentler process of maceration. Three week fermentation which includes extended maceration with malolactic fermentation occurring in barrels. A combination of wild yeast and cultured yeast is used.

Maturation: Older French oak for sixteen months. Some new oak is introduced but I prefer minimal oak character believing new oak can mask the fruitiness of the wine. Unfined and unfiltered.

Winemaker's Notes: Medium deep ruby. Nose of crushed ripe blackberries. More pure and less over-the-top than the '97. In the mouth similar flavors of predominately ripe berries. Moderately sweet ripe tannins, good acidity, good concentration, long finish. 364 cases bottled in March 2000.

1998 Mancini Ranch, (Mt. Olivet) Russian River

A new old-vine vineyard for us. It is located across the road from Zeigler Vineyard and was, like Zeigler, planted in the mid-1920's. The Mt. Olivet is an historic place name. I first noticed the name on an old wooden picking box printed with Frank Mancini's father's name followed by "Mt Olivet." I asked Frank if that had been the name of the winery his father had and he said no, it was the name of the district. Puzzled, I asked him where the mountain was (the vineyard is on the Santa Rosa plain, an area surrounding De Loach Vineyards and only slightly less flat than Kansas), and he replied, "the Holy Land". He went on to explain that the original settlers had called it Mt. Olivet but than it was later shortened simply to Olivet.

We took over the management of this vineyard, which is also the source of our Cotes du Rosa, in 1998. It was an expensive decision for us. Historically production has been in the 20 ton range but in 1998, due to El Nino (the year without a spring), it dropped to just over 8 tons. Farming costs greatly exceeded the value of the crop, but the fruit was quite concentrated and the resultant wine quite rich. Because of the uneven bloom that year, we went through the vineyard at verasion and dropped fruit that was either considerably ahead or behind the majority. At harvest we spent about twice the normal time picking. We went through on a vine-by-vine basis examining every cluster before they went into the picking lugs. Anything that showed signs of pink was left behind. In most cases the fruit had to be picked to examine it, since the outside clusters looked perfect but the backside would be pink. As a result the sugar levels were higher than we would have liked (no slightly under-ripe fruit to balance it), but the acid levels were quite high and there was almost no raising.

Composition: 100% Zinfandel

Vinification and Maturation remains virtually the same with all Zinfandels.

W i n e m a k e r ' s   N o t e s : Concentrated nose of very ripe, (not late harvest) fruit. Black berries, red plums, and black raspberries. In the mouth quite rich, concentrated flavors analogous to nose, with a hint of black pepper. Only moderate tannin. Good acid balance. It hides its considerable 16% plus alcohol well. The most "full bore" of this year's zins. Very tasty. If you like the "big boy" style of zin, this one's for you! 503 cases bottled in March 2000.

Joseph Swan Vineyards also produces Pinot Noir, Mourvèdre, Pinot Gris, Chardonnay, Cabernet Sauvignon and Rhône style blends.

Karly Vineyards

Karly Vineyards
11076 Bell Road
P.O. Box 729
Plymouth, CA 95669
Phone: 209-245-3922 Fax: 209-245-4874

Karly Wines was started from scratch by Lawrence L. (Buck) Cobb and wife Karly in 1980. The business has grown from a few hundred cases made in crude quarters to a modern facility turning out eleven thousand cases annually. Growth has been very controlled as Buck and Karly have always focused first on quality. The winery has long been known for Zinfandel and Sauvignon Blanc, but a new line of Rhone styled wines show great promise. About 70% of Karly output derives from estate vines; the balance from old Zinfandel vineyards owned by neighbors.

Proprietor: Buck and Karly Cobb   **Winemaker**: Buck Cobb

The Vineyards:

The climate and soil conditions in Amador County create grapes and wines of identifiable character, quite different from wines produced in the coastal area. In summer, Karly's rolling, canyon-side terrain catches the cool, southwesterly breezes that blow in from the Sacramento River delta during the afternoons. In late winter and spring, the canyons and draws act as natural drains for the freezing air. Thus, the harsh hot-cold extremes of the Sierra foothills are moderated at Karly, and frost is less of a problem. The soils; weathered granite, stream and glacial deposits eroded from the Sierra, are rich in minerals, dominated by iron, manganese and even trace amounts of gold. The texture is silty to rocky gravel with stone outbreaks varying from cobbles to room sized boulders. Clay occurs only occasionally in low areas. Water drainage is excellent - a feature common to all great vineyards.

Most of the vineyards at Karly were planted in 1981 and 1983, with small blocks added and grafted later. Following classic practice in the area, Zinfandel is dry farmed, head trained and spur pruned. Yield is about four tons to the acre.

Farming philosophy at Karly has always followed the theme of minimal intervention. Buck is also honing a system of integrated pest management that should, eventually, permit almost total organic farming.

1997 Zinfandel Warrior Fires

This is our fourth vintage of Warrior Fires Zinfandel. There are those wines that make themselves with minimal intervention; Warrior Fires is not one of them. The wine character is created by blending the two estate vineyards (one of which is on the fire-blackened earth for which the wine is named) are the backbone. A combination of technology, vigilance and plain old pigheadedness is needed to make this wine.

The 1997 vintage is the product of a long, moist winter and an unusually cool summer. Consequently, the vines enjoyed a leisurely growing season and a late harvest. These factors resulted in wines that attained complete ripeness with surprisingly strong acid tartness.

Appellation: Amador County
Composition: 80% Zinfandel and 20% Petite Sirah
Vinification and Maturation: After sixteen vintages there isn't much new to say about our production methods. We destem and crush into five-ton open top fermenters, hand agitate the mash of whole grapes for about a week while they ferment, drain and shovel the skins into our reliable Butcher tank-press, finish the primary fermentation in stainless uprights, and then remove the wine to our well experienced collection of French and American barrels for about a year before bottling.

Related information:
Alcohol: 15.5%	**Residual sugar**: Dry
Brix at Harvest: 25.5	**Harvest Date**: 9/20 to 9/30/97
Bottling Date: 7/29/99	

Winemaker's Notes:
The Shenandoah Plateau of Amador County is a land of red oxidized iron earth. The soil in our Zinfandel and Petite Sirah block however are quite black. When we first planted the vines, tilling always initiated a treasure hunt for stone arrow points, skinning tools, and grinding bowls; obvious indicators of habitation by earlier Americans. Later, an archeological survey for a water impoundment permit explained the black earth. The vineyard location had been an aboriginal hunting camp, and the soil was blackened by the embers of centuries of fires. Small wonder the vines from this block produce distinctive wine. The term "big Zin" is deserved. The bouquet is a clean and powerful combination of herbs and berries. Now the fun begins; dominant dark berry flavors supported by caramelly

Mexican chocolate undertones, with a finish that last until after the dishes are done. If alcohol bothers you, drink something else.

1998 Zinfandel Pokerville

Appellation: Amador County
Composition: Zinfandel, Black Muscat, Grenache Noir
Vinification and Maturation: Same approach for all Zinfandel. The 1998 Pokerville is bottled early for maximum fruitiness and styled for immediate consumption.

Related information:
Alcohol: 13.8

Winemaker's Notes:
If your in-laws are coming over and you want to relax and enjoy a glass of wine, then this is the wine. We call Pokerville Zin our local Beaujolais. Its light body, floral bouquet and spicy flavors lends itself to leisurely and introductory consumption.
The 1998 Pokerville Zinfandel is made in a drier style than its predecessors. Comprised primarily of young estate grown zinfandel Pokerville zestiness is enhanced by modest additions of Grenache Noir and Black Muscat.

1998 Sadie Upton Zinfandel

Appellation: Amador County
This has been our top Zinfandel for almost two decades. The wine is 100% from the vines planted by Sadie and her (then) husband John in 1922. Sadie lived until 1985 tending the grapes herself for six decades. We now do business with her son Lloyd.

The wine from this vineyard is always special. The thin to non-existent soil overlies granite slabs. The vineyard faces south collecting maximum sunshine, and is fully exposed to the westerly Delta breeze. The vines stress badly in dry years. Somehow all these factors combine to deliver what is almost always our best Zinfandel. When the magic does not happen, we blend it in the other Zins.

1998 was an El Nino year, cool and rainy long into spring and even summer. The old Sadie vines are open to sunshine and air, and it did not matter. However, the cool season sealed a lot of acid. Firm acidity is not what Amador is known for, and this makes the vintage unique. There is plenty of berry fruit and the flinty-mineral popourri that is the signature of the vineyard. The firm acidic backbone of this vintage promises a wine that will develop for years and reward cellaring. I usually recommend optimum drinking about three to five years after the vintage, but this wine I would want to save eight to ten years.

Winemaking followed our time honored practice of hand punching four times daily for ten days, and then transferring to well seasoned French oak barrels for fourteen months aging. The wine was then bottled without filtration and aged four months before formal release (some of our insiders know our bottling schedule and jump the gun against pain of severe admonition for drinking too soon).

Related information:
Alcohol: 15.1%   **Residual sugar**: Dry
Brix at Harvest: 25.5

1998 Buck's Ten Point Zinfandel

Appellation: Amador County
The name of the wine is a play on words. Buck Cobb is an avid hunter, often by necessity as the estate vineyards are afflicted with a voracious resident deer population, who despite game fencing, seem to grow and breed at the same rate the freezer is emptied. Ten pointers are as rare in this part of California as ten-point wine, so we deemed it appropriate to name the wine for the particular vineyard and memorialize the buck that formerly ate, slept and entertained ladies in it.

1998 is the second Ten Point Zin. The first 1997 edition caught us totally by surprise, selling out quickly where as we had intended it to carry us through an entire year. It was the most successful wine in our history. Obviously we did something right, so we patterned the 1998 wine as close as possible to the first version, although no two wines are ever completely alike. The grapes used to make the wine are a blend from our Ten-Point and Buley vineyards. Wine from the Buley vineyard shows the strawberry end of the Zinfandel taste spectrum whereas the Ten Point vineyard has the dark berry rustic earthiness of the central Shenandoah plain. The blend of the two makes broad-shouldered richly fruited satisfying wine. This 1998 wine has a structure that should result in some age-ability.

Related information:
Alcohol: 15.1%   **Residual sugar**: Dry
Brix at Harvest: 25.   **Harvest Date**: 10/16 to 10/17/98

C e l l a r i n g   N o t e s :
While our Zins are often pleasant to drink ten years and more, I like them best three to five years after harvest.

Karly also produces Sauvignon Blanc, Syrah, Marsanne and experimental size lots of Grenache, Mourvèdre, Barbera and Roussanne

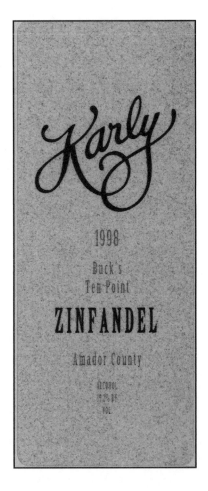

Kendall-Jackson

Kendall-Jackson
425 Aviation Blvd..
Santa Rosa, CA 95403
Phone: 707-544-4000 Fax: 707-569-0105
Email: mail@kjmail.com or www.kj.com

Kendall-Jackson's winemaking philosophy is to select fruit from the state's best vineyards and blend those lots to create a wine that can be relied on to deliver consistent quality and intense flavor from vintage to vintage.
Tasting room located at the Kendall-Jackscon Wine Center 5007 Fulton Road, Fulton. Open 10 a.m. to 5 p.m. daily. Self guided Sensory Garden Tours, wine tasting, special events and sales.
Owner: Jess Jackson **Winemaker**: Randy Ullom

The Vineyards:
Comprised of old vineyards in Redwood Valley, Ukiah, Dry Creek, and Amador county. Most of the vineyards are head pruned and dry farmed which produce deep color and the intense jammy spicy flavors that is characteristic of Zinfandel.

The Vintage:
1998 is the Zinfandel vintage of the decade. A cool spring, nice Indian summer and late harvest combined to produce a vintage of great depth and intensity of color, flavor, and structure.

1998 Vintners Reserve Zinfandel

Appellation: California
Composition: 99.1% Zinfandel, with a hint of Merlot and Petite Sirah
Vinification and Maturation: Grapes harvested at flavor maturity which translate into a harvest Brix of 23.5 to 25.0. The hand picked fruit was destemmed and crushed into chilled tanks. A 24 hour cold soak was followed with yeast inoculation and fermentation to dryness. The must was pumped over several times daily. Shortly after dryness the free run wine was separated from the skins and the skins pressed. Depending upon the wine, a portion of the press fraction may be added back to the

free run to increase structure. The wine was inoculated with malolactic bacteria and put to barrel. Better oak integration is achieved by barrel fermentation. Upon completion of malolactic the wine is sulfated to preserve fruit flavors. Barrel aged for 14 months in approximately 20% new French and American oak. The wines are racked or topped quarterly as appropriate for each lot for clarification and tannin maturation.

Related information:

Alcohol: 13.9%   **Residual sugar**: 0.07%

Brix at Harvest:25   **Harvest Date**: 10/01 -10/29/98

Production: 21,750 cases

Retail: $12.99

Winemaker's Notes:

Many fingernails were bitten and hairs lost by grape growers and winemakers waiting for the harvest of 1998 to happen. Fine autumn weather insured full ripeness which, for Zinfandel, translates into high alcohol and the occasional slow to finish fermentation. But '98's Zinfandel paid us back with a rare vintage in which all of our vineyards produced wine of such depth, character, body, and flavor that each one warranted bottling on its own. The blend turned out to be a wonderful synergism of everything Zinfandel. This wine is a sumptuous combination of cherry, blackberry, and raspberry jams with black pepper and clove spices. The wine's dark color promises the richness and body realized in the mouth. Structured for aging or consuming. WOW.

Kenwood Vineyards

Kenwood Vineyards
9592 Sonoma Highway
Kenwood, CA 95452
Phone: 707-833-5891 Fax: 707-833-6572
www. kenwoodvineyards.com

Kenwood Vineyards was founded in 1970 when wine enthusiasts from the San Francisco Bay area acquired the historic Pagani Brothers Winery (circa 1906) in Kenwood. True to its history, the winery continues to be housed in the original Sonoma barn buildings. Ideally located in Sonoma Valley, Kenwood Vineyards has from its first vintage been committed to producing premium varietal wines that reflect Sonoma County's finest vineyards in their character and style.

Tasting Room is open daily from 10 - 4:30 p.m..
Proprietor: Gary Heck
Winemaker: Michael Lee

Winemaking Philosophy: At Kenwood Vineyards, grapes from Sonoma County's best vineyards get the attention they deserve. The harvest from each vineyard is handled separately within the winery to preserve its individuality. Such small lot winemaking allows the winemaker to bring each lot of wine to its fullest potential.

1997 Mazzoni Zinfandel, Geyserville

Vineyard: Mazzoni Ranch is located in northern Sonoma County's Alexander Valley in the town of Geyserville. The climate of this appellation is known for producing very distinctive, fruit-forward Zinfandels. These growing conditions combined with cordon pruning in this 7-acre vineyard, promote an even, early ripening of the fruit. Kenwood has consistently found Mazzoni's grapes to be exceptional and several years ago decided to bottle it as a single vineyard designated wine.

Composition: 100% Zinfandel
Vinification and Maturation: Following a very good growing season in 1997, the Mazzoni crop was harvested at optimum ripeness in mid-

September. After fermentation in stainless steel tanks, the wine was aged in new American oak for 15 months.

Related information:
  Alcohol: 13.5%   **Bottling Date**:6/99
  Production: 692 cases   **Retail**:$20.00

W i n e m a k e r ' s N o t e s : Brimming with ripe plum and black cherry fruit flavors, the Mazzoni Ranch Zinfandel is a richly layered, full-bodied wine with toasty nuances on the finish. The delicious supple flavors complement pasta, veal, lamb and steak.

1997 Jack London Vineyard Zinfandel

Vineyard: Long recognized for its superior red wine crop, the Jack London Vineyard produced its first Zinfandel in 1988. The climate conditions of this viticultural area, located on Sonoma Mountain, give a southeasterly exposure creating complexity and balance promoted by its long, cool growing season. The resulting fruitier characteristics, and the influence of the vineyards red volcanic soil, produce an excellent, very distinctive Zinfandel crop.
Composition: 100% Zinfandel
Vinification and Maturation: Harvested in early October, this wine is fermented in stainless steel tanks, and then is aged in American and French oak barrels for approximately 14 months.

Related information:
Alcohol: 13.9%   **Bottling Date**:4/99
Production: 8330 cases   **Retail**: $20.00

W i n e m a k e r ' s N o t e s : Vibrant wild berry and cherry flavors accented by a hint of cedar and spice meld into an elegant, smooth finish. The wild berry and black cherry flavors of this medium-bodied Zinfandel are a perfect complement to rich tomato sauce dishes, grilled fish and steak, and rich chocolate desserts.

1997 Upper Weise Ranch

Vineyard: Planted over 70 years ago, the Upper Weise Ranch is an organically farmed Zinfandel vineyard. The vineyard sits at an elevation of 1,000 feet located on the Mayacamas Range above Sonoma Valley. The head-trained vines grow in non-irrigated red volcanic soil that yields a desirably low 2 ½ to 3 tons per acre.

Composition: 100% Zinfandel

Vinification and Maturation: Farmed by Kenwood since 1971, in 1994 the winery chose to reserve a portion of the Upper Weise crop for a single vineyard bottling because of its distinctive, old-vine qualities. After an excellent growing season in 1997, the grapes were harvested in September. The wine is fermented in stainless steel tanks, and aged in French and American Oak barrels for 16 months.

Related information:
 Alcohol: 13.9%   **Bottling Date**: 5/98
 Production: 1935 cases   **Retail**: $20.00

W i n e m a k e r ' s   N o t e s : Spicy peppercorn, clove and berry fruit flavors are well balanced in this supple medium-bodied, dry Zinfandel that finishes with depth and elegance. The spicy flavors of the Upper Weise Zinfandel pair well with barbecued meats and rich tomato sauce dishes.

1997 Sonoma Valley Zinfandel

Vineyard: The 1997 vintage is blended from vineyards throughout Sonoma County. A portion of the blend comes from three old-vine vineyards - Upper Weise, Ivy Glen and Bald Mountain. The balance of the blend comes from vineyards ranging in age from 15-40 years.

Composition: 98% Zinfandel 2% Petite Sirah

Vinification and Maturation: The 1997 growing season was warm with an early spring bloom, allowing longer hang time for the grapes which were harvested in mid-September. Each vineyard lot was fermented separately in stainless steel and aged individually in American and French Oak barrels for 14 months.

Related information:
 Alcohol: 13.8%   **Bottling Date**: 7/99
 Production: 7000 cases   **Retail**: $15.00

W i n e m a k e r ' s   N o t e s : Medium-bodied with black cherry, spice and vanilla oak flavors that finish with classic peppery Zinfandel notes. Our Sonoma Valley Zinfandel is delicious with barbecued meats, rich tomato based dishes, and spicy ethnic cuisine.

1994 Nuns Canyon Zinfandel

Vineyard: Nuns Canyon , Sonoma Valley. Located at an elevation of 1,200 feet on the Mayacamas Range, the 8 acre Nuns Canyon Vineyard yields a desirably low 2 ½ to 3 tons per acre of intensely flavored fruit.
Composition: 100% Zinfandel
Vinification and Maturation: The growing season of 1997 provided excellent conditions in Sonoma Valley for red wine crops. After harvest in mid-September, the Zinfandel was fermented in stainless steel tanks and aged in French and American oak for 16 months.

Related information:
  Alcohol: 13.5%   **Bottling Date**: 6/99
  Production:1,178   **Retail**: $20.00

W i n e m a k e r ' s   N o t e s : Robust ripe cherry, wild raspberry and plum fruit flavors and aromas are accented by nuances of black pepper, clove and spice. This is a firm and supple, classic Zinfandel. The supple flavors of the Nuns Canyon Zinfandel complements roasted and grilled meat dishes, hearty pastas and stews, and spicy ethnic cuisine.

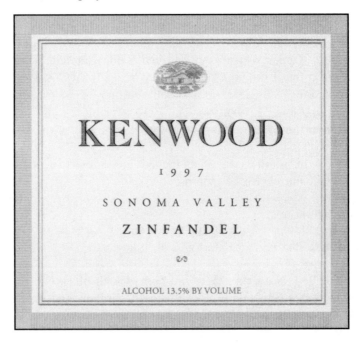

Kunde Estate Winery

Kunde Estate Winery
10155 Sonoma Highway
Kenwood, CA 95452
Phone: 707-833-5501 Fax: 707-833-2204 Website http://www.kunde.com

Since the early 1900's, four generations of the Kunde family have preserved the family tradition of farming their fertile, highly praised vineyards located in the heart of the Sonoma Valley. Their 2,000 acre estate, of which 750 acres have been planted to vineyards, boasts such acclaimed designations as Wildwood, Kinneybrook and Bell ranches. Their winery and 1/2 mile of aging caves mark the family's commitment to producing world class,100% estate grown varietals. Year after year these wines have earned Kunde the reputation of consistently producing premium quality vintage dated varietals.

Tasting Room open daily 11:00 to 4:00. Cave Tours available Friday, Saturday and Sunday. Wine Club.

Proprietor: The Kunde Family   **Winemaker**: David Noyes

The Vineyard:

The vineyards of the Kunde Estate rise from about 400 feet elevation at the valley floor to over 1200 feet at the upper reaches of the ranch. All contain greater or lesser amounts of clay and volcanic rock. Distinctive and unique, our red soils provide a striking contrast to the deep green of the vine's summer foliage. Any Kunde will proudly tell you that the Estate contains 85% of the 'Red Clay #2' soil series found in Sonoma County. Our exposures run from South-West to North-West catching the last rays of the setting sun, but also, during harvest, a cool wind that comes down from Santa Rosa and the Russian River Valley.

Alluvial soils, made up of clay, dirt and stones washed down from the hills above, support the valley floor vineyards. Most of these soils run only 4 to 8 feet deep, ending at a clay hardpan that stops root growth and reduces vigor in these cooler sites. Vine roots go deeper in the hillside soils - we found roots 30 to 40 feet deep as we drilled our caves.

These deep, clay-rich, soils are only moderately fertile but they hold water well and allow for fairly vigorous vine growth through the middle of the season. This pushes our harvest dates later into the fall, and allows our fruit to mature at slightly cooler temperatures than dryer, warmer soils would. This in turn helps account for the deep color and rich fruit aromatics that characterize our wines.

We grow Zinfandel at all elevations on the ranch. Later maturing, cooler, valley floor blocks contribute color and aroma, while a south facing, earlier ripening, hillside block at about 1000 feet adds tannin and briar components. Our Century Vines block lies somewhere between these extremes at about 600 to 800 feet elevation lining both sides of a small winter creek in deep red soils. The varying exposures and elevations within this 30 acre piece lead to several picking dates to ensure every grape is harvested at peak maturity. These ancient head-trained vines, planted in narrow terraces suitable for horse-drawn cultivation speak mutely, yet eloquently, to the history of grape-growing in Sonoma Valley.

The Vintage:
Despite record high global temperatures in 1998, we would call 1998 a cool, late season for the Sonoma Valley. A wet spring delayed bud break and hampered flowering, leading to small crops of intensely flavored fruit. The late season had most growers worried…would fall rains ruin the already meager crop? Fortunately the rains held off and we were able to pick all our fruit at good maturity. The cool fall weather during the last stages of ripening preserved and enhanced fruit and spice aromatics in all varietals, including Zinfandel. Overall the 1998 wines show great complexity of fruit and spice flavors and a lively range to tannin effects; peppery in youth turning elegant as the wines develop.

1998 Zinfandel Kunde Estates

Appellation: Sonoma Valley
Vineyard Designation: Kunde Vineyards.
Composition: 86% Zinfandel, 9% Petite Sirah, 2% Grenache, 2% Mourvèdre, 1% Syrah
Vinification and Maturation: We vinify all of our Zins using the same traditional, hand-crafted approach. All our grapes are hand harvested. We de-stem, but do not crush the fruit, then allow a two to three day cold maceration, for color extraction, before the native yeasts begin the fermentation. We pump over frequently and press at dryness or after several days of maceration, when we judge the color tannin balance to be the best. Aged for 12 months in new to 3 year old cooperage, both American and Czech oak, 30% new, medium plus toast.

Related information:

Alcohol: 13.9   **Residual sugar**: dry
Brix at Harvest: 25.5
Harvest Date: For the Zin, 10/7-10/16, other varieties 11/3
Bottling Date: 8/99   **Release Date**: 10/99
Production: 8,000 cases sold at winery and distributed.
Retail: $14.00

W i n e m a k e r ' s   N o t e s :
Well before Prohibition, Zinfandel delivered some of the best of California's wines. Using a traditional hand-crafted approach we produce our Estate Zinfandel to capture all the excitement and rich fruit flavor Zinfandel has to offer.

Our 1998 Estate Zinfandel shows a higher level of fruit and herbal aromatics than typical. A long cool fall season enhanced these more volatile components, creating a delightfully flavorful wine. The flavors play against a backdrop of ripe fruit and sweet vanilla flavors.

1998 Century Vines Zinfandel
Kunde Estates

Appellation: Sonoma Valley
Vineyard Designation: Kunde Vineyards.
Composition: 90% Zinfandel, 10% mixed blacks, mainly Alicante Bouschet and Petite Sirah
Vinification and Maturation: We hand harvest the Zinfandel grapes for this wine from our 118 year old Century Vines block. De-stemmed but not crushed, the must undergoes a two to three day cool maceration for color extraction before the native yeast begin the fermentation. We pump over frequently at the peak of fermentation, and press at dryness or after several days of extended maceration, when we judge the color tannin balance to be the best.Aged for 14 months in new to 3 year old cooperage, both American and Czech oak, 30% new, medium plus toast.

Related information:
Alcohol: 14.6   **Residual sugar**: dry
Brix at Harvest: 26.5   **Harvest Date**: October 10th and 16th
Bottling Date: 4/00   **Release Date**: August 2000
Production: 1,500 cases
Retail: $22.00

Winemaker's Notes:
Our Century Vines Zinfandel represents the best this 118 year old vineyard has to offer. These ancient vines speak to Zinfandel's place in the history of wine growing in the Sonoma Valley, and to our pride in maintaining this tradition. We aim to showcase the unique character of this vineyard through the use of traditional, hand-crafted winemaking techniques. The sweet vanilla of both European and American oak deftly complements the full flavored, slightly briary, Zinfandel fruit.

Bigger and bolder than the Estate Zinfandel, our 1998 Century Vines Zinfandel displays more cocoa, brown sugar, or maple syrup flavors. The character of the 1998 vintage shows through, too, in the bright fruit and herb aromatics. I sometimes think of raspberries and chocolate with a touch of anise as I attempt to describe the flavors we find from these 100 year old vines.

1998 Robusto Zinfandel

Appellation: Sonoma Valley
Vineyard Designation: Kunde Vineyards.
Composition: 100% Zinfandel
Vinification and Maturation: We harvest Robusto grapes at peak maturity, from the highest and rockiest section of our 118 year old Zinfandel vineyard. Over 26° Brix at picking, the must is slow to start fermentation and may take three weeks or even longer to finish. Otherwise, we use the same techniques as for our other Zinfandel wines. Aged for 14 months in new to three year old cooperage, both American and Czech oak, 30% new, medium plus toast.

Related information:
Alcohol: 15.8   **Residual sugar**: dry
Brix at Harvest: 30.0   **Harvest Date**: October 11th
Bottling Date: 4/00   **Release Date**: September 2000
Production: 500 cases sold at winery and distributed.
Retail: $30.00

Winemaker's Notes:
Left to ripen beyond normal maturity levels, the Zinfandel grape can make wines like Robusto. These are extreme wines, with concentrated fruit and high alcohol

levels made for the special occasion, not for every day. Dry, concentrated wines such as Robusto find their counterpart in Italian Amarones, where harvested grapes are left to dry in covered lofts until they reach the sugar levels that Zinfandel can achieve on the vine in California.

Our 'Robusto' originates from the most stressed, lowest cropped vines of the Century Vines block. Located at the top of the hillside in the rockiest soil of the block, these ancient vines may only bear two or three clusters of fruit a year. Even more concentrated than the Century Vines, the flavors shift even more towards cocoa, chocolate and dried fruits. True to the vintage, however, a persistent herbal note links and complements the ripe fruit.

Kunde Estate Winery also produces Magnolia Lane Sauvignon Blanc, Viognier, Chardonnay, Kinneybrook Chardonnay, Wildwood Chardonnay, Reserve Chardonnay, Syrah, Merlot, Cabernet Sauvignon and Reserve Cabernet Sauvignon

Kunin Wines

Kunin Wines
c/o Central Coast Wine Services
2717 Aviation Way
Santa Maria, CA 93455
Phone: 805-689-3545 Fax: 805-884-1513
Email: seth@kuninwines.com

Kunin Wines is a small, boutique operation based in Santa Barbara County, and is owned by the Kunin family (Norman, Rebecca, Seth and Christopher). 1998 was the first commercial vintage for Kunin and consisted of two zinfandel bottlings - one, a blend of two Paso Robles vineyards called "Westside," the other, a single vineyard designated wine from Dante Dusi's vineyard, also in Paso Robles. For 1999 and beyond, we have added syrah and viognier to our portfolio.

Winemaking Philosophy:

We make wine that is based on fruit. "Isn't all wine based on fruit?" you might ask. Well, in so far that all wine comes from grapes, yes it is. However, some wines are treated in ways that either mask the fruit, or let the vibrant, jammy flavors escape - often replaced by more tart, dried out fruit flavors, if any at all. If you lose the youthful, ripe fruit components before the wine goes into the bottle, you can never get them back. Some wines don't even get the chance to have big, bold fruit flavors because the grapes were farmed improperly or not picked at the optimum ripeness, while others go too far, and pick overripe fruit, yielding wines that may be off-dry and age awkwardly.

We strive to source our grapes from vineyards that pay close attention to the condition and balance of the vines, and monitor the ripening process closely to ensure that the grapes are picked at the peak of their maturity. Once we have gotten the fruit flavors to the winery, we destem, ferment in open-top bins with some whole berries, punch down by hand, and press gently to barrel. No more than 30% of the barrels are new, to prevent oak flavors from dominating the fruit. The wines are barrel aged for less than 12 months, to avoid both oak flavor domination and the "drying out" of their fruit components. Filtration is avoided whenever possible. After bottling, the wine is aged for a minimum of 3 months before release to let the flavors resolve themselves. We bottle vineyard designated wines whenever feasible so that the unique flavor profiles of each particular vineyard's fruit can be showcased by themselves, rather than muddled by blending.

1997 Zinfandel, Paso Robles, "Westside"

Vineyard: Templeton Hills and Dante Dusi Vineyards

Related information:
Alcohol: 14.1%
Release Date: January 2000
Production: 115 cases
Retail: $18.00

1997 Zinfandel, Paso Robles, "Dante Dusi Vineyards"

Vineyard: Dante Dusi Vineyards

Related information:
Alcohol: 15.4%
Release Date: January 2000
Production: 110 cases
Retail: $26.00

Winemaker's Notes:
Dark plumy color. Aromas of ripe dark fruits - blackberry, plum and cassis with some earth/spice notes. Big and rich on the palate, with a long finish of blackberry jam. Moderate tannins. Drinkable now, but will continue to develop over the next 1-3 years.

La Crema

La Crema
8075 Martinelli Road
Forestville, CA 95436
Phone: 707-571-1504 or 800-588-5298
Fax: 707-517-1448
Email: www.lacrema.com or mail@lacrema.com

Established in 1979, La Crema's focus is in Pinot Noir, Chardonnay and Zinfandel. Located in the heart of the Russian River Valley, La Crema works with many unique vineyards throughout Northern California's Cold Coast. With its close proximity to the Pacific Ocean and consequent cold growing conditions, the quality of fruit from this region is outstanding. Winemaker Jeff Stewart uses traditional techniques and small lot methods to craft rich, velvety textured wines that showcase these small distinctive vineyards.

Tours and tasting by appointment only.

Owners: Jennifer Jackson Hartford and Laura Jackson Giron

Winemaker: Jeff Stewart

Winemaking Philosophy:

At La Crema, our winemaking style is dictated by the fruit that we source from the cooler climates of Sonoma County. Working with fruit of high fruit intensity, bright acidity and terroir driven complexity, our goal is to balance these characteristics by hand harvesting, hand sorting, and gentle handling during fermentation. This allows the terroir of each vineyard site to shine through and creates wines with seamless intensity and complexity.

1998 Sonoma Coast Zinfandel

Appellation: Sonoma Coast

Composition: 100% Zinfandel

Vineyard: 94% Sonoma Coast, 64% Russian River which included six old vine vineyards. This blend of cool climate Zinfandel vineyards gives the 1998 Sonoma Coast Zinfandel its distinct forward fruit aromas and flavors, along with bright acidity.

Vinification and Maturation: Hand harvested fruit, gently destemmed and pumped to open top fermenters. Five day cold soak followed by 12

day fermentation, with three punch downs per day. Aged 10 months in French Oak barrels, 35% new.

Related information:

Alcohol: 14.5%   **Residual sugar**: Dry
Harvest Date: October 20-29th
Bottling Date: August 99   **Release Date**: September 2000
Production: 2450 cases sold through the winery and distributed
Retail: $18.99

Winemaker's Notes:

1998 was marked by a cool, wet spring, followed by a mild summer. This led to a long growing season which gave the fruit long hang time and great flavor maturity. The 1998 La Crema Zinfandel shows characteristics consistent with the region the fruit is sourced from: bright raspberry, blackberry and chocolate in the aroma, and black cherry, bright acidity and a long finish with soft tannins and flavors of black pepper.

Cellar Notes:

Previous vintages of La Crema Zinfandel are developing nicely with age. The 1995 (our first vintage) is very rich and complex today, with fresh fruit still driving the wine. All of our Zinfandels have the ability to age gracefully because of the acid/tannin backbone present in Russian River Valley and cooler climate fruit.

La Crema Winery also produces Pinot Noir and Chardonnay.

Lamborn Family Vineyards

Lamborn Family Wine Company LLC
120 Village Square Suite 13
Orinda, CA 94563
Phone: 925-254-0511   Fax: 925-254-4531
email: mike@lamborn.com
Web site: www.lamborn.com

Tasting Room:
Napa Wine Company
7830-40 St. Helena Hwy.
Oakville, CA 94562
Phone: 707-944-1710
Mailing list. Mail order.
Proprietors and Grape growers: Michael and Terry Lamborn
Winemaker: Heidi Peterson Barrett

Winemaking Philosophy :
As it is with many of the world's greatest works of art, balance is the essence of their greatness. Heidi's winemaking approach follows this very philosophy. "I prefer to utilize a combination of traditional and modern techniques in my winemaking. One of my goals is to maximize the unique characteristics of each estate vineyard, guiding the wine to achieve its highest potential while always focusing on quality and balance."

The Vineyard:
Robert Lamborn, founder, and his wife Janet, sold their zinfandel farm in 1998 retiring to St. Helena where growing "friendships" has been their main crop...production has been plentiful!

Son Michael, and his wife Terry, along with their two grown sons Matt and Brian, continue the family tradition of zinfandel on Howell Mountain. Their five-acre vineyard was planted in 1981 with the identical nursery material used by Mike's father in 1979. Deep red volcanic loam soils and dry farming encourage this 2,200 ft elevation vineyard to produce intensely flavored grapes.

Another significant influence upon the fruit of this tiny sub-appellation is temperature. During the growing season, the average daytime temperature is ten degrees cooler than Napa Valley and ten degrees warmer at night. This particular vineyard site lies 125 feet below the ridge of Howell Mountain on the Eastern side. The slight elevation increase to the west protects the vineyard from the customary afternoon winds which blow from the Pacific Ocean. As expected, the Eastern side of Howell Mountain receives a larger amount of rain than its Western counterpart. However, because the Lamborn's five acre vineyard occupies the top of a promontory which on three sides drops off at a 60% slope to Pope Valley 1600 feet below, the soil drainage is superior. The same terrain which benefits grapevine growth is very limiting to vineyard development and size. The last ground flat enough to plant will be developed to three acres of Cabernet in 2001.

Vineyards on Howell Mountain normally experience bud break later than in the valley and ripen later as well. In 1998 the Lamborns re-trellised the vineyard to a split canopy management system. Customizing a concept developed by Dr. Richard Smart of Australia, the new trellis showed significant promise in 1999 by ripening, on average, a week earlier than the neighboring vineyards and yielding a crop or record quality and quantity (3 tons per acre).

Mother Nature is always the farmer's number one challenge. Many valley growers are plagued before harvest by migrating flocks of birds that can consume large quantities of fruit. Lamborn Family Vineyards has never experienced bird problems...but we have our annual bear visit. Mt. St. Helena is home to a sizable bear population who love a "sweet" grape treat before the winter sets in.

1998 Vintage:
The "El Nino" weather phenomenon was the dominant force in Northern California in 1998, bring unseasonably late rains into Man and June, delaying all stages of vine growth and development by a month. Generally, cluster counts were normal but cluster weights were very low die to poor berry set. 1998 growers reported very mixed results based upon their grape variety and vineyard location. Again, because Howell Mountain begins the growing season later that the valley, we were spared the "beating" some growers experienced. After a late start, the season continued to be a challenge due to cooler than normal temperatures. Leaf removal and crop thinning were "key" management tools for this vintage. At Lamborn Family Vineyards we ended up removing 50% of the crop to assure ripening and quality; a smaller vintage resulted producing only

980 cases. We were blessed with a late warming trend that lasted well into October and "saved the day". Many wines from the 1998 vintage, Lamborn included, will prove to be exceptional; however, one would be hard pressed to find a grape grower who would want to repeat the experience.

1998 Zinfandel "The Obsidian Effect"

Obsidian, named for its supposed discover, Obsius, is a natural glass formed deep beneath the earth's surface, super heated and discharged by volcanic activity. The grapes of Howell Mountain are indeed affected by the mountain's historic volcanic beginnings. These influences are many: the altitude which greatly impacts the temperatures, the scarcity of water, and sunlight exposure. The soils, their mineral content, drainage, and depth, all result from the volcano.

Appellation: Howell Mountain **Vineyard Designation**: Lamborn Family Vineyards, with a small percentage coming from Black-Sears Ranch and Pringle Family Vineyards.
Composition: 100% Zinfandel
Vinification and Maturation :
Our 1998 vintage was harvested October 16th at 23.9 Brix. The wine was aged in 20% new French made American oak barrels. The 1998 vintage was bottled November 23, 1999, with an anticipated release of October 2000. Each year since 1989 we have named the wine after an event or significant influence. The 1998 vintage, "The Obsidian Effect", recognizes Howell Mountain's volcanic beginnings and the consequential impact on the grapes grown there.

Winemaker's Notes:

The tasting notes for this vintage are as of March 2000. The wine will make remarkable changes between now and release.
A dense ruby-purple wine with exceptional balance. Minerals and black fruit are revealed in the bouquet. This is a full-bodied wine with a long, nutty/toasty finish. The has solid "Howell Mountain" structure and with identifiable vineyard characteristics including the pepper/spice component.

Cellar Notes:

1994 "Queens Vintage" The hallmark of Howell Mountain Zinfandel, the berry characteristics are holding very well. The tannins have softened considerably but the wine still remains structured. This wine is at its consumption apex.
1995 "French Connection" The intensity of the fruit remains in the forefront while the French Oak flavors develop more taffy/caramel characteristics. Good tannin forms the

framework around the highly extracted fruit. This wine will continue to improve and soften over one to two years.

1996 "Family Connection" This vintage is reminiscent of the 1989 when the intense heat of the growing season produced a zinfandel lighter in style. Also, like the 1989, this vintage was extremely harsh on the pallet for months following release. Age has worked it's magic and this wine is drinking wonderfully. As if they came from nowhere, the fruit and peppery spices are a pleasure upon the pallet. The backbone of this vintage will allow it t cellar and improve for another three to four years.

1997 "The Team Connection" A textbook growing year produced a textbook crop. This, the first vintage from winemaker Heidi Barrett, reflects the perfect balance and great elegance. Deep color, silky texture, dark fruit and spice, with a long, deep , toasty finish. It will continue to evolve and last for 7-8 years.

1999 yet to be named...1999 truly appears to be the vintage of the decade. Mother Nature could not have provided a better vineyard year - it was a classic! The combination of a new trellis design, changes in both the pruning approach and canopy management, resulted in the most exciting zinfandel crop in twenty years. Barrel samples continue to support our enthusiasm. Approximately 1300 cases produced.

Lamborn Family Vineyards produces only Zinfandel! A late harvest is made when weather is appropriate. A port was made in 1996.

Winemaker Heidi Peterson Barrett also produces wine under her own label La Sirena.

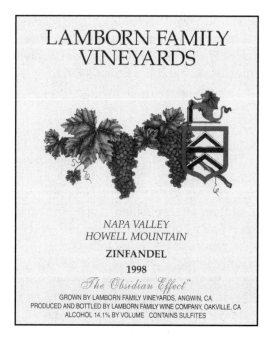

LAMBORN FAMILY
VINEYARDS

NAPA VALLEY
HOWELL MOUNTAIN
ZINFANDEL
1998
The Obsidian Effect
GROWN BY LAMBORN FAMILY VINEYARDS, ANGWIN, CA
PRODUCED AND BOTTLED BY LAMBORN FAMILY WINE COMPANY, OAKVILLE, CA
ALCOHOL 14.1% BY VOLUME   CONTAINS SULFITES

Lolonis Vineyards & Winery

Lolonis Vineyards & Winery
1905 Road D
Redwood Valley, CA 95470
Mailing address: 1904 Olympic Blvd., Suite 8A
Walnut Creek, CA 94596
Phone: 925-518-8482 Fax: 925-938-8069
Email: lolonis@pacbell.net

Enotria is Greek for "Land of the Vines". That was what Tryfon Lolonis saw when he arrived in California from Greece in 1914. At the age of sixteen, Tryfon surveyed Mendocino's Redwood Valley, where he noticed the similarity to his native Velherna, Greece. He cleared the land by hand and built a house for his bride-to-be, Eugenia whom he had met in Greece. She arrived in San Francisco in 1920 to be with Tryfon. They married and raised ten children. As the family grew, so did the vineyards.

In the late 1940's Tryfon and Eugenia's son, Nick, was among the first students to study viticulture and enology at the University of California, Davis. His classmate was Andre Tchelistcheff's son, Dimitri. Field trips for the students meant a trip to Beaulieu Vineyards to study with Tchelistcheff. When Nick decided to study for a Masters degree, his professors were Amerine, Winkler, and Olmo. Since Nick was one of the few students whose family owned a vineyard, the U.C. professors asked if the Lolonis family was interested in field-testing new French vine clones. The original clones used for the U.C. Davis test still exist.

After Tryfon Lolonis passed away in 1961, son Petros, Nick and Ulysses assumed vineyard management. Their dedication to quality is evident in the grapes that are harvested from Lolonis vineyards. No chemical pesticides are used. Instead ladybugs are spread out twice each growing season to eliminate pests. To further insure organic purity, vine leaves are picked off by hand to expose grapes to sun and aid air circulation. Each acre produces only three to four tons of grapes per year. This commitment to quality is unique to have spanned a period of forty years and continues today.

The Lolonis family is proud of their heritage. They are the only Greek grape growers who own their own vineyards and whose parents planted the original vines that still exit today. Their motivation remains that of sharing - the love of land, the love of family and the love of great wines.

President: Petros Lolonis   **Winemaker**: Kevin Blundell

Vice-President Sales: Maureen O'Reilly-Lolonis
Vice-President Marketing: Phillip Lolonis
Consulting Enologist: Jed T. Steele

1997 Zinfandel Private Reserve

The 1997 was an exceptional year for Redwood Valley Zinfandel. It has an intense varietal character, consisting of ripe raspberries, blackberries and sweet oak. We keep the Reserve in the barrel two months longer than our Estate bottling. We use French and some American Oak. We only produce 900 cases and the wine should be quite big but maintain elegance.

Organically grown on 8x14 spacing, the vines are 50 years old with head trained and spur pruned trellising. Ladybugs and praying mantises are used to combat pests and leaf pulling and cover crops (vetch and legumes) are also practiced.

Lolonis wines have a nice amount of fruit and spice. The wine is also well balanced and has a round finish. The fruit comes from the age of the vines, and climate, which gets hot in midday and cools down at night. The spice comes from the soil. 3 to 4 feet of soil with many different types throughout the vineyard. Clay and loam from sedimentary rock, that has a deep hue of red, ferric soil. The well balance nice finish comes from our winemaker, Kevin Blundell and Consultant, Jed Steele.

Related information:
Alcohol: 13.8%
TA: 0.55
pH: 3.64
Bottling Date: June 1999
Production: 900 cases
Retail: $27.00
Lolonis also produced a 1997 Estate Zinfandel that retails for $18.00

The Vineyards:
Lolonis Vineyards began practicing organic farming since 1956, when our brother, Nick, graduated from the University of California at Davis, with a master's degree in Viticulture and Enology. No pesticides are used: Instead, ladybugs are used to eliminate pests such as leaf hoppers and red spiders, while praying mantis enjoy a feast of lace wings which cause damage to young grape

leaves. The following are some differences between Organic and Non-organic vineyard cultural practices.

Non-organic:
To combat the pathogens of grapes, non-organic growers use pesticides and or herbicides. Today many growers are aware of the potential hazards of pesticide use and are going toward a farming method called "Sustainable Agriculture". It relies much less on pesticides but is not called "Organic" because the use of pesticides is lower but not extinct.

Organic:
Culturally, about the only material applied to the vineyard is sulfur, which is organically approved because it is naturally mined. It is applied to prevent mildew from forming and is used about three times a year, beginning when the new growth is about six inches long. The second application is just before bloom and the third shortly after completion of the bloom period.

When weeds are growing in the area around and between each vine, called the berm, they can out compete young vines and are a habitat for mice, voles and gophers. To rid this area of weeds we use a "Kimco" cultivator instead of herbicides. The "Kimco" cultivator is mounted on the tractor and moves in and out of the berm with a spring action.

To help prevent mold from developing on the grapes (especially the Chardonnay), we pull most of the leaves off one side of each vine. This is done by hand and it allows indirect sunlight for more fruit intensity and better aids wind movement within the vine canopy, controlling mold and mildew better and using chemicals.

In the early fall we plant a cover crop (vetches of legumes) in every other row. Cover crops produce nitrogen as a fertilizer, biomass to improve soil porosity and provide a habitat for beneficial insects such as Ladybugs.

Lolonis Winery also produces Chardonnay, Cabernet Sauvignon, Carignane, Gamay, Merlot, Petite Sirah, Pinot Noir , and Sauvignon Blanc.

Lonetree Winery

Lonetree Winery
P.O. Box 401
Philo, CA 95466
Phone and Fax: 888-686-9463 Net: CHZin @aol.com

Lonetree Winery is a partnership between sparkling wine pioneer John Scharffenberger and viticulturist Casey Hartlip. Using fruit grown exclusively from the Eaglepoint Ranch, one of Mendocino County's highest elevation vineyards, they have combined their efforts to produce some of California's finest red wines. Lonetree takes winemaking as seriously as their farming. The goal: make the most powerful, concentrated wines without losing sight of balance. Zinfandel, Syrah, and Sangiovese each made in their own unique styles. Lonetree wines, from the land to the bottle, experience, dedication and quality. Mail order.

Proprietors and **Winemakers**: John Scharffenberger and Casey Hartlip

Winemaking Philosophy:
We are a grower-owner winery and strive to bring the essential characteristics of each vintage to the bottle. Lonetree takes winemaking as seriously as their farming. The goal: make the most powerful, concentrated wines without losing sight of balance. Zinfandel, Syrah, and Sangiovese, each made in their own unique styles. Lonetree wines, from the land to the bottle, experience, dedication and quality.

The Vineyard:
As with all Lonetree wines, the grapes for the Lonetree Zinfandel came from the Eaglepoint Ranch, located at 1800 ft. elevation on the eastern side of the Ukiah Valley. More than just a vineyard, Eaglepoint is more like an experiment in viticulture techniques and range management. The 74 acre vineyard sits in the middle of nearly three square miles of upland range and timberland. 29 separate blocks, 6 different varieties, numerous exposures and trellis systems, all at very high elevation make for very intense farming. Probably the most important ingredient is the soil. Moderately fertile decomposed sandstone is the primary feature with numerous veins of "timber red" loams as well as upland pasture soil types. Like many of the finest vineyards of Europe, the vines have a more challenged existence. Crop load is kept at moderate levels either naturally or by manual thinning, yielding fruit with extreme concentration and intense varietal

character. John's dedication to the land isn't just practiced in the vineyard but extends to the entire property. Reforestation, soil conservation and preserving native habitat are more than projects, they're a way of life. When the work in the vineyard slows, you'll find our crews thinning conifers, removing fallen and damaged hardwoods and planting fir, cedar and big leaf maples. Soil erosion is a huge concern for mountain vineyards, our crews have built siltation dams, installed countless rolls of drain tile and prepare the vineyard for each winter's rains.

1998 Lonetree Zinfandel

Appellation: Mendocino

Vineyard Designation: Eaglepoint Ranch. This vineyard, with its shallow mountain-top soils and western exposure, consistently produces grapes of great concentration and depth. The fruit for this wine came from our original planting of head-trained vines which were planted in 1975 on St. George rootstock.

Composition: 95% Zinfandel and 5% Petite Sirah.

Vinification and Maturation: At first daylight, the grapes are hand harvested and the fruit crushed into a small tank with 20% whole clusters. Cold soaked for three days at 55°F the must was inoculated with a cultured Rhône yeast. The wine was pumped over twice a day and was encouraged to reach a peak fermentation temperature of 88°F. At approximately 3 Brix the wine was pressed gently and returned to tank to finish fermentation. At dryness, the wine was racked to a combination of 25% new Tonnellerie Radoux American oak and 75% 3-5 year old French oak barrels. Over the 11 months in barrel, the wine was racked two more times. Before bottling, a fining of 2 egg-whites per barrel was performed to soften the tannins.

Related information:

Alcohol: 14.8%	**Residual sugar**: Less than 0.1%
Brix at Harvest: 25.8	**Harvest Date**: 10/17/98
TA: .58	**pH**: 3.78
Production: 598 cases distributed.	

The Vintage:

1998 will be known by many as the 'El Niño' vintage. As farmers, this was one of the most difficult years (sharing that distinction with '83, '89, and '93). Over 70 inches of rain fell on our vineyard that year. The soil was wet and the vines got off to a slow start. It was a cool summer for the most part with only one warm spell in late July.

We have over 20 years of vineyard and weather data for our site. Usually verasion begins about the 18th of July, in '98 our first berries changed color on August 17th! We knew that we were in for trouble. We drop all second crop and pull leaves on the east side every year regardless of weather. But we knew the most important thing we could do in that '98 vintage was to reduce our crop size by thinning.

The fall weather in September was not much help to us either. Daytime temperatures were much below normal and ripening was slow. The lucky thing was that any damaging rain held off. We picked the last of our Zinfandel on October 29th but we got it up to 24.5 Brix! This was the latest that we've ever harvested Zinfandel in the history of our vineyard. Talk about hang time!

Winemaker's Notes:

We didn't make a '97 Zin because our mountain climate gave us a short crop and we had commitments to our long standing winery customers. Knowing that, we made the 1998 wine in a more approachable style. It's amazing how supple it is already. If it's a Zinfandel and it's from Eaglepoint, you know what to expect...FRUIT, FRUIT, FRUIT! A juicy mixture of ripe blackberry and boysenberry.

Food pairing: Grilled lamb, tomato based pastas.

Lonetree supplies grapes to Hidden Cellars, Edmeades, Rosenblum, Navarro, Claudia Springs, Sean Thakery, Copain, Fieldstone, Pepi, Parducci and Surh -Luchtel.

Loxton

Loxton Cellars
P.O. Box 70
Glen Ellen, CA 95442
Phone and Fax: 707-833-6766
Website: www.loxtonwines.com

Loxton Cellars was founded in 1996 focusing on producing just two wines, Syrah and Zinfandel. These two wines reflect the passions of the owner/winemaker, an Australian brought up on a Shiraz vineyard in South Australia, but now living in California where Zinfandel is the adopted grape.

Winery is not open to the public. Sales via mailing list, selected restaurants and fine wine stores.

Owner and Winemaker: Chris Loxton

Winemaking Philosophy:
The Loxton Zinfandels are made to showcase the varietal and vineyard site in which the grapes are grown. I seek to make wines of balance; the goal being to have fruit driven wines of complexity and depth of character without being heavy or unbalanced. The approach is deliberately low tech and non-interventionist. The wines are made without adding commercial yeasts. Grapes and wine are handled gently with the aim of neither fining nor filtering where possible.

1998 Priest Ranch/Orion Vineyard, Napa Valley

Appellation: Napa Valley
Vineyard: Priest Ranch/Orion Vineyard
The old Priest Ranch is located in the mountains east of Saint Helena near Chiles Valley. The Zinfandel vineyard is about thirty years old and under its new ownership is dry farmed and totally organic. The commitment is to growing the best fruit possible. Attention to vineyard, work, and respect for the environment will produce nothing less.
Composition: 100% Zinfandel
Vinification and Maturation: The wine is made in open top fermenters using native yeasts and the traditional hand punch down of juice and skins to extract color and tannins. A few days cold soak precedes

fermentation as this helps gain extraction. After pressing, the almost finished fermenting juice is pumped into new to five year old French oak barrels to finish fermentation and malolactic. Minimal racking was done until the wine was bottled unfined and minimally filtered.

Related information:

Alcohol: 13.7%   **Residual sugar**: Dry
Harvest Date: Mid to late October 1998
Bottling Date: August, 1999
Release Date: May, 2000
Production: 370 cases sold through mailing list and limited distribution.
Retail: $16.00

Winemaker's Notes:

Despite a wet spring and cool weather, the decision to farm organically and the requisite careful attention to the vineyards paid off well in 1998. Harvest approached slowly and we waited and waited into October to pick. Warm and sunny weather finally pushed the maturation of flavors and tannins to desired levels and we started hand picking after first sending in a crew to discard fruit not fully ripened. As a result, we have a wine showing tons of cherry fruit, some pepper, good acid balance, soft tannins and just a whisper of oak. Balanced, full of fruit, medium bodied, it shows the best qualities of the '98 vintage.

Food pairing: BBQ chicken, pork with cherry sauce, pasta with tomato and Portobello mushrooms.

Cellar Notes:

1997 Napa Valley: Soft and ripe, medium bodied, black cherry fruit and a hint of pepper, the wine has opened up considerably in the last 6 months and is now approaching its optimum drinking.

1996 Russian River Valley: Lighter vintage, shows wonderful complexity of raspberry and pepper fruit and has improved in body since bottling. Fully ready, it should be enjoyed now and into 2001.

Marietta Cellars

Marietta Cellars
P.O. Box 1436
Healdsburg, CA 95448
Phone: 707-433-2747 Fax: 707-857-4910

Marietta Cellars is located just outside the little town of Geyserville in California's premier wine growing region, Sonoma County. Owner and winemaker Chris Bilbro founded Marietta Cellars in 1979 after cutting his wines business teeth by working at Bandiera Winery, founded by his grandparents Emil and Luduina Bandiera in 1937.

Marietta Cellars was named for Marietta Pardini, Grandfather Emils' sister-in-law. She was a native of Tuscany, Italy, and emigrated to the United States with her mother at the age of ten. She was a wonderful cook; known for her large garden and love of hunting wild mushrooms. A favorite saying was: "Most anyone can make a good meal with lots of ingredients, it is when you can make good things with very little that you have talent."

As Chris was growing up Marietta taught him how to put flavors together, how to learn to use and trust his palate and sense of taste, and develop a style. For Chris, this educational experience was so valuable he showed his appreciation by naming the winery after her.

Because of the limited production and availability of the wines, there are no tours, tasting or retail sales on the premises.

Proprietor and Winemaker: Chris Bilbro

Winemaking Philosophy:

As winemaker, Chris describes his wines as "full bodied, fleshy reds with lots of fruit and structure." While they are created to age well for a decade or two, one can enjoy them in their youth as well. Marietta Cellars' style combines basic winemaking skills with the best fruit available. The final product is well-made and mildly seasoned, with an abundance of fruit characteristics and only a touch of oak. While all the wines are aged in 60 gallon French and American cooperage, Chris' winemaking style emphasizes the unique attributes of the varietals rather than the oak. A little bit of blending is used to take full advantage of flavor and combination for the optimum in taste and enjoyment.

The wine of Marietta Cellars are a personification of Chris Bilbros' personality, outlook on life and general lifestyle. Chris believes that good wine is the perfect complement to good food and good friends.

1997 Zinfandel Sonoma County

Appellation: Sonoma County
Vineyards: 100% Sonoma County grapes from vineyards in Geyserville, Healdsburg, Russian River and Forestville.
Composition: 95% Zinfandel and 5% Petite Sirah.
Vinification and Maturation: Stainless steel fermentation with pump overs. Fourteen days to dryness then pressed. 15 months in French oak barrels.

Related information:
Alcohol: 14.9%
Brix at Harvest: 23.5 - 24.5
Release Date: 9/01/99
Production: 6,800 cases

Winemaker's Notes:
This newest vintage of Marietta Cellars' Zinfandel is so deep, dark, and black purple in color that it is almost opaque. Equally intense is the aroma which is heady with camphor, black pepper and blackberry. The viscous, thick-bodied mouthfeel coats the palate with blackberry and raspberry flavors and just a hint of oak and black pepper. The tannins are soft, yet, noticeable, and allow the mixture of flavors to cling to the palate and produce an enjoyable long, lingering finish. A touch of Petite Sirah was added to the Zinfandel for firmness and structure. As with all Marietta Zinfandels, the 1997 may be enjoyed today for the Maximum in Big Zin style or allowed to gracefully mellow for up to eight years without losing its lush character.

Marietta Cellars also produces Syrah, Cabernet Sauvignon, Old Red Vine - a proprietary red wine and Angels Cuvée - a blend of Zinfandel Petite Sirah, Syrah and a bit of Carignane.

Mayo Family Winery

Mayo Family Winery
9200 Sonoma Highway
Kenwood, CA 95452
Phone: 707-833-3300 Fax: 707-833-2883
Email: jeffmayo@aol.com
Website: www.mayofamilywinery.com

Mayo Family Winery was founded in 1993 not necessarily by design, but by divine intervention and a new found passion. Henry and his wife, Diane, purchased property in Sonoma County in 1984, built their dream house and moved in with their three children in 1987. In 1990 Henry underwent heart bypass surgery and during convalescence decided to change the pace of his life. Diane and Henry had become passionate about wine and thought grape farming would bring them in closer touch with their new found pleasure. The intent was to grow grapes and sell them but the connection they felt for their vineyard transcended into the desire to make wine with the fruit they grew. The commitment to vineyard, winemaking and passion for life continues as the Mayo Family celebrates their 7th year as one of Sonoma County's highly regarded producers.
Family Wineries of Sonoma Tasting Room, Sonoma Highway, Kenwood.
Tasting room open daily.
Owners: Harry and Diane Mayo   **Winemaker**: Chris Stanton

1998 Ricci Vineyard Zinfandel

Appellation: Russian River Valley
Vineyard: This wine represents the first vintage from the Mayo Family Winery's new Ricci Vineyard. The vineyard is located in the warmer, easternmost reaches of the Russian River Valley north of the town of Healdsburg on Limerick Lane. The vines are 30-40 years old.
Composition: 100% Zinfandel
Vinification and Maturation: Unfiltered, tank fermented, pump-overs. 12 months in French and American oak.

Related information:
Alcohol: 14.8%
pH: 3.49

Total Acidity: .69
Release Date: November, 1999
Production: 249 cases
Retail: $25.00

W i n e m a k e r ' s   N o t e s :
An intense Zinfandel reflecting the age of the vines and limited cropping, the Ricci Vineyard Zinfandel brings deep garnet colors to the table. The nose is perfumed with aromas of red raspberry, violets, fresh jam with hints of sage and anise. On the palate the intensity of the wine shows best with mouthfilling flavors, a medium to large structure and an impressive finish.

1998 Ricci Vineyard Zinfandel, Reserve

Related information:
Alcohol: 15.5%   **pH**: 3.57
Total Acidity: .66
Release Date: November, 1999
Production: 215 cases   **Retail**: $30.00

W i n e m a k e r ' s   N o t e s :
This wine is a profoundly intense rendition of California Zinfandel. Grapes from the ripest areas of the vineyard, blocks 2, 3, and 6, were used in the wine. The bottling is limited to free-run juice. Winemaker Chris Stanton chose an equal combination of mostly new French and American barrels for the 12 months of oak aging. The wine was bottle unfiltered and unfined. The result is everything the family hoped from this remarkable vineyard.

A deep purple colors the wine's appearance. The nose is intensely perfumed with brooding blackberry and bramblebush aromas jumping out of the glass. Cinnamon, raisin and prune aromas envelope the dark fruit. On entrance, the wine is huge over the palate leaving more blackberry flavors and prominent rosemary and licorice highlight. This is the type of hedonistic wine that gives Zinfandel its good name.

Mayo Family Winery also produces Chardonnay, Pinot Noir, Cabernet Sauvignon and Merlot.

Meeker Vineyard

The Meeker Vineyard
21035 Geyserville Ave
P.O. Box 215
Healdsburg, CA 95448
Phone: 707-431-2148 Fax: 707-431-2549

When Billy Idol sang "with a rebel yell, she cried more, more, more" he could very well have been singing about Meeker wines. Or maybe it was the old tasting room…The Teepee, that inspired similar hedonistic cries from visitors to Meeker Vineyards. Long time fans of Meeker know that teepee well, their voices cloaked in bitter-sweet melancholy whenever they speak of it. But life is full of change and the tasting room moved from its rustic digs to another unassuming site: a bank in downtown Geyserville, just a short drive up Hwy 101 from Healdsburg. Plans for Meekers tasting room are ever evolving with a possibility of a teepee once again rising as a beacon to all those who love wine for the pure joy of it. That is what Meeker wines are all about…love of life, joy, and celebration… just give me more, more, more.

Tasting room open, call for hours

Owners: Charlie and Molly Meeker   **Winemaker**: John Allen Burtner

1997 Gold Leaf Cuvee Zinfandel

The '97 Gold Leaf Cuvee offers a provocative nose of raspberry, cream, deep dish berry cobbler, Asian spices and new (15% American, 25% French) oak. These flavors are echoed on the palate as layer after layer of fruit unfold during its lengthy and memorable finish.

Vineyard: Dry Creek Valley

Composition: 100% Zinfandel

Vinification and Maturation: The fruit for this release came from four different sites on the west side of Dry Creek Valley. The vines range from babies (20 years old) to mature (93 years old) and produced remarkable ripeness and fruit intensity. After a slow, cool fermentation and extended maceration, the lots of wine were kept separate for 12 months of barrel aging. The lots were than selected, blended and allowed to "marry" for 5 months before bottling.

Related information:
Meeker Vineyard, Sonoma County   174

Alcohol: 14.8%
pH: 3.51
Production: 1452 cases

1997 8th Rack Zinfandel "Whatta Rack"

This is a 100% Zinfandel from two of the very best Zinfandel growing appellations within Sonoma County. The nose is redolent of sour cherry, mulberry, dark chocolate and cola with tasty new oak aromatics. On the palate, there are cherry/berry fruit and dark chocolate - a good acid structure with a rounded mouth feel and smoky oak tannins emerging in the lengthy finish. This "Rack" Zinfandel can certainly age 2-3 years but its accessibility cries out for immediate consumption.

Related information:
Alcohol: 13.9%
pH: 3.49
TA: .71
Production: 986 cases

Miner Family Vineyards

Miner Family Vineyards
P.O. Box 367
7850 Silverado Trail
Oakville, CA 94562
Phone: 707-944-9500 Fax: 707-945-1280
www.minerwines.com

Founded in 1998 by Dave Miner, President of Oakville Ranch Vineyards, along with his wife Emily and his parents, Ed and Norma, Miner Family Vineyards produces small lot, hand crafted wines using fruit from Oakville Ranch, Ed and Norma's vineyard and other carefully selected California vineyards. Miner Family Vineyards is located at 7850 Silverado Trail in Oakville. The state of the art winery has approximately 100,000 case capacity and 20,000 square feet of newly constructed caves for barrel aging and storage. We produce and sell both Miner Family Vineyards and Oakville Ranch wines here, as well as producing wines for many clients as a custom crush operation.
Owners: Dave and Emily Miner, Ed and Norma Miner
Winemaker: Gary Brookman

Winemaking Philosophy:
We produce limited quantities of Cabernet Sauvignon, Merlot, Cabernet Franc, Zinfandel, Sangiovese, Pinot Noir, Chardonnay, Viognier and Sauvignon Blanc using a combination of old world winemaking techniques and modern technology. The focus of our label is to produce high end, reserve style wines that reflect the unique characteristics of individual vineyards or terroir where specific varietals grow best.

1998 Miner Family Vineyards Zinfandel

Appellation: Napa
Vineyard: All the Zinfandel for the '98 vintage comes from a single vineyard in Calistoga. We work closely with our grower while grapes mature refining vineyard practices such as crop and leaf thinning. We feel the warmer temperatures in Calistoga are ideal for maturing Zinfandel in Napa.

Vintage: The 1998 growing season and harvest can be summed up in two words...long and small. After a freak hailstorm in April cut crop by 40-45%, mild summer temperatures delayed harvest by about a month. 1998 was a difficult year to get Zinfandel as ripe as it needs for real "Zin" character but we left ours out as long as possible, as you can see by the 25° harvest Brix.

Composition: 95% Zinfandel 5% Sangiovese

Vinification and Maturation: Hand harvest and delivered in ½ ton bins. Crushed to stainless steel tank, cold soaked for 24-48 hours with gentle pump overs, then inoculated. Total skin contact 14 days with minimal pump over after 9 days. Full malolactic fermentation. Aged in 50% new American oak barrels for 14 months. No fining.

Related information:

Alcohol: 14.4% **Residual sugar**: Dry
Brix at harvest: 25
Harvest Date: October 9th, 1998
Bottling Date: January, 2000
Release Date: May 1st, 2000
Production: 568 cases sold at winery and through distribution
Retail: $22.00

Winemaker's Notes:

The grapes for this Zinfandel were harvested from a single vineyard in Calistoga and blended with small amounts of Sangiovese to add brightness and complexity. Our '98 Zinfandel is full of ripe berry and spice notes with a hint of dried fruit and a touch of leather.

Cellar Notes:

1997 was our first vintage. Pepper, spice, and cassis. Structure holding nicely with good balance, oak integration and a long finish. Drink now through 2002.

Montevina

Montevina Winery
20680 Shenandoah School Road
Plymouth, CA 95669
Phone: 209-245-6942 Fax: 209-245-6617 Web site http://www.montevina.com
Email: QandA@montevina.com

After purchasing Montevina Winery in 1988, the Trinchero family began expanding the winery's Shenandoah Valley estate vineyard. Reflecting Amador's Mediterranean viticultural heritage, climate and topography, the Trincheros committed the majority of this new acreage to classic Italian red grape varieties.

This year our new wine making facility will be completed. The expansion includes a larger space for fermentation, storage and blending and a crush pad. A new lab, office space and warehouse are all housed under one roof. One of the bigger expansions is the Oak room. Oak storage plays a huge role in aging wine. Oak not only adds flavor, it enhances the aging process. In our old winery we had room for only 600 barrels, but the 22,000 square feet of barrel space in our new facility, we have capacity for more than 6,000 barrels. Their addition will result in more complex, flavorful wines than we've produced in the past. Complementing the barrel room is an oak tank room. Our oak upright room is where a portion of the vintage will be aged prior to blending and bottling. In these large oak tanks (the smallest is 1,000 gallons and the largest is 6,100 gallons), we're not trying to add flavor to the wine as we are with barrels. Rather, our goal is simply to let the wine age and gain complexity.

We are very excited about the new facility for all it's challenges and possibilities with the end result being the production of higher-quality Montevina wines.
Tasting Room and picnic area open daily from 11:00 am to 4:00 pm. Wine Club. Newsletter. Special Events.
Proprietor: The Trinchero Family   **Winemaker**: Jeff Meyers

Winemaking Philosophy:
True to their Italian heritage (the Trincheros' parents emigrated to the U.S. in the 1920's from Italy's Piedmont district) the Trinchero believe wine is a part of life. Their focus is on Italian red varieties and they bring a dedication and determination to producing the best Amador County wine possible.

1998 Brioso

Vinification and Maturation:
Brioso (Italian for "a lively tune") is the lightest of Montevina's four Zinfandels. The grapes come from our estate Shenandoah Valley vineyard and neighboring Amador County vineyards.

Half the zin grapes selected for Brioso undergo a standard yeast fermentation, while the balance are fermented by carbonic maceration.
In this technique, which the French use to make Beaujolais, whole, uncrushed grape clusters are dumped into a tank, which is sealed. The weight of the clusters breaks the fruit at the bottom of the tank, releasing juice which is inoculated with a cultured yeast to initiate conventional fermentation. The carbon dioxide created by the yeast fermentation blankets the whole clusters at the top of the tank.

In this anaerobic (oxygen-free) setting, a partial, enzymatic (rather than yeast) fermentation transpires within the individual, uncrushed berries, producing a small amount of alcohol. The unique chemistry of this intracellular fermentation extracts volatile aroma and flavor compounds from the unbroken grape skins (compound minimized in conventional fermentations), imparting a singular quality to the wine.
The result is a softer, fruitier red that is lighter in color, tannin and acid, but possess intensely fruity aromas and juicy, red fruit flavors - in short, a classic Beaujolais-style wine.

Related information:
Alcohol: 12.9%   **Residual sugar**: Dry
TA: 0.5 gm/100ml   **pH**: 3.49
Bottling Date: 5-99   **Release Date**: 7-00
Production: 8,700.
Retail: $7.00

Winemaker's Notes:
The 1998 vintage of Brioso display a lovely, translucent cherry color and fresh, fruity, slightly spicy, berry-like aromas immediately reminiscent of Beaujolais. The medium-bodied palate offers juicy, sappy, strawberry and raspberry flavors that are clean, crisp and beautifully balanced, with soft tannins and a hint of varietal pepper spice. This engaging red wine, best served slightly chilled, is a perfect for novice and red wine drinkers as well as those seeking a lighter red for picnics, patio sipping, barbecues, and other casual meals. It beautifully

complements grilled salmon, roast chicken and fettucine alfredo, among other Mediterranean-style dishes.

1998 Amador County Zinfandel

Vinification and Maturation:
The 1998 vintage in California produced big, ripe, high-alcohol zinfandels. This was true, as well, in Amador County, resulting in the most intense Montevina Zinfandel bottling in years, one reminiscent of the winery's house style during the 1970's.

After saignée, (removing 15% of the free run juice after crushing to concentrate flavors), the zinfandel juice was fermented in stainless steel tanks for ten days at an average 80°F. Subsequent to gentle tank pressing and natural clarification, a blend of free run and press wines was aged for an average of six months in French-coopered American oak barrel and puncheons and lightly filtered before bottling.

Related information:
Alcohol: 15.0%   **Residual sugar**: Dry
TA: 0.60 gm/100ml   **pH**: 3.53
Bottling Date: 1-2000   **Release Date**: 9-2000
Production: 25000.
Retail: $9.95

Winemaker's Notes:
Our 1998 Zinfandel exhibits a super-ripe aroma of raisined berry fruit and tremendously rich, extracted, mouthfilling flavors. Despite its high alcohol, it is remarkably well-balanced with surprisingly soft tannins and excellent acidity. This is a terrifically robust red wine to pair with grilled steaks, barbecued ribs, hearty, spicy pastas and stews, and flavorful cheeses like aged sharp cheddars, bleu, and Stilton.

1998 Terra d'Oro Zinfandel

Vinification and Maturation:
Montevina's Terra d'Oro Zinfandel is vinified from low yielding, (2-4 tons/acre) 46 year old vines grown at the winery's 380-acre, sustainably farmed, estate vineyard in the Shenandoah Valley of Amador County. The grapes are harvested at optimum ripeness (23.5°-25° Brix) . After crushing, 15% of the free run juice is drawn off (saignée), to maximize

the skins-to-juice ratio and extract greater color, tannin and flavor. The wine ferments at approximately 75°F for 10 to 12 days to fully extract color and flavor while softening tannins. At dryness, it is pressed in state of the art tank presses, minimizing rough tannins from the grape skins. The Terra d'Oro Zinfandel then ages for up to a year in a combination of new and once-filled American oak barrels. Prior to bottling, it is lightly filtered to insure stability.

The 1998 vintage produced extremely ripe, concentrated red wines in Amador County. Grapes for the Terra d'Oro Zinfandel were harvested in late September at an average 25° Brix. After saignée, the juice was fermented using the rack-and-return method practiced in Bordeaux (draining the tank of juice and then re-circulating it over the fallen cap of skins and seeds to enhance color and flavor extraction), and kept in contact with the skins for 12 days. After pressing, the new wine was settled and racked to American oak barrels, 40% new, where it aged for 14 months before being lightly filtered and bottled in May, 2000.

Related information:
Alcohol: 15.5%   **Residual sugar**: Dry
TA: 0.62 gm/100ml   **pH**: 3.50
Bottling Date: 5-00   **Release Date**: 1-01
Production: 2200   **Retail**: $16.00

W i n e m a k e r ' s   N o t e s :
The 1998 Terra d'Oro Zinfandel exhibits a dark, saturated color and a complex aroma of ripe raspberry fruit enhanced by pronounced notes of spice, anise and toasty oak. The wine has a rich, round texture with intense, spicy-berry flavors that also carry hints of cedar and tobacco. It is beautifully balanced with a firm tannic structure for long-term aging. This superb Amador County Zinfandel can be enjoyed now or over the next decade with hearty foods and flavorful cheeses, or simply savored on its own.

1998 Terra d'Oro Zinfandel,
Deaver Ranch Vineyard

Vinification and Maturation: Beginning in 1968, and for nearly 30 years thereafter, Sutter Home Winery in Napa Valley produced intense, robust zinfandels from a vineyard planted in 1881 in the Shenandoah Valley of Amador County. In fact, Sutter Home's Deaver Ranch Zinfandel was one of the first wines to convince wine lovers that the zinfandel grape could produce world-class red wines. In 1997, Montevina Winery, (which was purchased in 1988 by the Trinchero family who have owned Sutter Home since 1947), began receiving the old-vine Deaver Ranch fruit. Bottled under Montevina's Terra d'Oro reserve wine label, this is the first vineyard-designation wine Montevina has produced since its inception in 1970.

The 1998 vintage produced extremely ripe and concentrated red wines in Amador County. Grapes for the Deaver Ranch zin were harvested in late September at 25.2° Brix. After saignée, the juice was fermented using the rack-and-return method and kept in contact with the skins for 14 days. After pressing, the new wine was settled and racked to American oak barrels, 40% new, where it aged for 14 months before being lightly filtered and bottled in April, 2000.

Related information:

Alcohol: 15.5%	**Residual sugar**: Dry
TA: 0.68 gm/100ml	**pH**: 3.44
Bottling Date: 5-00	**Release Date**: 1-01
Production: 2700	
Retail: $22.00	

Winemaker's Notes:

This magnificent, old-vine zinfandel should delight aficionados of "old-style" Amador County zins. Nearly black in color, it offers an immense aroma of super-ripe blackberry and black raspberry fruit, mingled with distinct scents of raisins, cocoa, briary spice, and oak vanillin. The ripe berry and milk chocolate flavors are lush and mouthfilling, and, despite the wine's high alcohol, which is barely noticeable, it is beautifully balanced, with a very soft texture, unobtrusive tannins, and excellent acidity, which materializes in a lip-smacking finish. This is a truly extraordinary, memorable zinfandel, which can be enjoyed now or over the next decade with the heartiest foods, or simply savored on its own.

Montevina Winery also produces Sangiovese, Barbera, Aleatico, Refosco, Nebbiolo Rosato, Syrah, Fume Blanc, and White Zinfandel. Limited quantities of Montevina's Italian style Terra D'Oror (Land of Gold). Other proprietary blends include Brioso, Montanaro and Matrimonio.

Morgan Winery

Morgan Winery
590 Brunken Ave.
Salinas, CA 93901
Phone: 831-751-7777 Fax: 831-751-7780
website: morganwinery.com

    With seventeen successful vintages behind them, Dan and Donna Lee, owners of Morgan Winery, are now able to take the winery to the next level of quality. With the purchase of 65 acres in the up and coming Santa Lucia Highlands appellation, 50 acres have been planted to Chardonnay and Pinot Noir. Considered to be the "Cote d'Or" of Monterey County, these benchland vineyards are in a prime location for the production of cool weather varietals. Notable neighbors include the vineyards of Talbott and Mer Soleil. The first crop will be harvested in the fall of 1999. A new winery will be constructed, with groundbreaking to take place in the summer of 2000.
Tours and tasting are conducted by the winery staff-an appointment is required.
Proprietors: Daniel Morgan Lee and Donna Lee
Winemaker: Dean DeKorth

Winemaking Philosophy:
Our focus in on allowing the individual qualities of each vineyard to dictate the style of the wine produced.

1998 Morgan Zinfandel

    Appellation: Dry Creek Valley
    Vineyard: Sbragia Vineyard
    Composition: Small amounts (5-10% each) of Petite Sirah and Carignane are added for structure and complexity.
    Vinification and Maturation: The grapes were hand sorted at the crusher. All of the grape clusters were destemmed. The crusher's rollers were set to allow approximately 25% whole berries through. No commercial yeasts were used. Primary fermentation and extraction took place in a combination of small, open top vats-employing manual punchdowns, and a medium size closed top tank-using a gentle sprinkler

system. After fermentation, there was a period of 5-7 days skin contact for the closed top tank. Secondary (malolactic) fermentation occurred in barrels. The wine was then aged for approximately 16 months, with periodic racking (as needed), in new to four year old French oak barrels. The final cuvee was assembled after 11 months, and then returned to barrels for 5 additional months to ensure integration. None of the red wines at Morgan receive fining or filtration prior to bottling.

Related information:

Alcohol: 14.6%   **Residual sugar**: Dry
Brix at Harvest:24.8   **Harvest Date**: October 8th
Bottling Date: Feb 2000
Production: 1350 cases distributed and sold at the winery.
Retail: $18.00

Winemaker's Notes:

The 1998 Zinfandel has an opaque ruby color, typical bright berry-ish fruit tones, vanillin and spice barrel notes, and a positive yet silky mouthfeel. In keeping with our desired style, an excellent balance of characters -fruit/oak/acidity/- are in evidence.

Morgan Winery also produces Chardonnay, Pinot Noir, Syrah, Pinot Gris and Sauvignon Blanc.

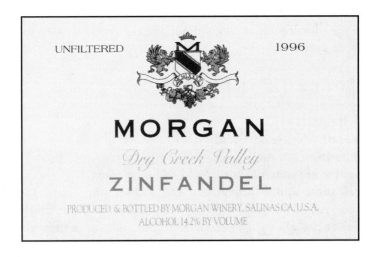

Murphy-Goode Estate Winery

Murphy - Goode Estate Winery
4001 Highway 128
P.O. Box 158
Geyserville, CA 95441
Phone: 707-431-7644 Fax: 707-431-8640

As a winery founded by grape growers, the emphasis in all our wines is on fruit - the flavor of the grape. With Zinfandel, we have selected vineyards in Alexander Valley and Dry Creek Valley that can produce wines with lots of raspberry and cherry flavors.

Tasting room opened daily 10:30 - 4:30. No tours

Proprietors: Tim Murphy, Dale Goode and Dave Ready

Winemaker: Christina Benz

1998 Sonoma County Zinfandel

Appellation: Sonoma County

Vineyard Designation: The 1998 comes from three Alexander Valley vineyards, all on sloped ground. The vineyards are 15 years old or older. 1998 was characterized by lower than average crop levels and a late spring and cool September. Rigorous cluster thinning was done to insure ripeness in a short season.

Composition: 100% Zinfandel

Vinification and Maturation: After cold soaking overnight, fermentation is started and the tank is allowed to warm. Extraction is done by pumping over 3 times per day at the start of fermentation, slowing to once a day at the finish. The wine remained on skins for 12-16 days. The 1998 vintage spent 13 months in French and American oak of which 37% are new.

Related information:
Alcohol: 14.5%
Residual sugar: Dry
Harvest Date: 10/10-10/16/98
Bottling Date: 3/6/00
Production: 5,800

Winemaker's Notes:

Patience was needed in the vineyard. The grapes got a late start and seemed to take forever to reach just the level of ripeness we require. With Zinfandel this is more of a look and a taste than an actual measurement.: looking for full coloration with a slight amount of shrivel and a "gobs of fruit" taste without rasiny or cooked flavors.

C e l l a r   N o t e s : Cellar notes were compiled March 15, 2000 from wines in Tim Murphy's private library.

1993 Cuneo & Saini Vineyard, Dry Creek Valley
Our first Zinfandel, produced from what we determined were "middle-aged" vines (40 to 60 years old). The aromas are spicy with a lot of sweet vanilla oak. Still a bright mouthful with ripe plum flavors. The tannins have smoothed out completely. Drink now or over the next 2 years.

1994
None left in Tim's library…must have been good!

1995 Zinfandel, 57% Dry Creek Valley and 43% Alexander Valley
Brick red in color. Displaying toffee and white pepper aromas with jammy undertones. Soft and supple, finishing with flavors of currants. Drink now.

1996 Sonoma County
Garnet red. Complex aromas - black fruit, spice, briar, tomato leaf. Round and full-bodied with tart fruit on the finish. Will go another 5 years.

1997 Sonoma County, Liar's Dice
Deep red. Very aromatic and spicy with strawberry jam aromas. Dense fruit. Nice measure of acid and tannin. Finishes with sweet, concentrated fruit. Definitely a "buy", but okay to hold for 5 years.

Nalle Winery

Nalle Winery
P.O. Box 454
Healdsburg, CA 95448
Phone: 707-433-1040 Fax: 707-433-6062
e-mail: dbnalle@sonic.net

Located in the Zinfandel Center of the Universe, a.k.a. Dry Creek Valley, is the winery of Nalle, specializing in Zinfandel since 1984. The winery building was completed in 1990 on property that is not just home to Nalle Zin, but to Doug, his wife and partner Lee, and their two children, Andy and Sam.
No tasting room. Mailing list.
Proprietors: Doug and Lee Nalle   **Winemaker**: Doug Nalle

Winemaking Philosophy:
Nalle Zinfandel is produced with methods used world wide for fine red wine: hand sorted grapes, open top tank fermentation, French oak cooperage, frequent racking and egg white fining. Our goal is to express, in the wine's youth, the intrinsic "zinberry" fruit and elegance of Dry Creek Valley Zinfandel, and with bottle age, complex characteristics that define all world class wines.

1998 Nalle Zinfandel

Appellation: Dry Creek Valley
Vineyards: I work very closely with four outstanding grape growers in Dry Creek Valley to produce the highest quality grapes.   Old   vines, averaging 55 years, on low yielding bench land soils include 8-10% Petite Sirah and 2-4% other (Carignane, Syrah, Alicante Bouschet and Gamay).
Composition: Zinfandel with field blend which adds color, middle body and tannin.
Vinification and Maturation: Hand sorted at the crusher. 1,250 gallon open top tanks with 7-15 days of skin contact. Gentle pump overs as to not agitate the fruit. Malolactic fermentation. Aging in new to four year old French oak barrels, frequent racking, and egg white fining.

Related information:
Alcohol: 13.4%   **Residual sugar**: Dry
Brix at Harvest:23.2
Harvest Date: 9-19 to 10-17-98
Bottling Date: August 1999   **Release Date**: Spring 2000
Production: 2,100 cases distributed and sold through the mailing list.
Retail: $25.00

Winemaker's Notes:

Late, later and finally wet. What a wacky harvest. One could almost hear "Woe is me" drifting across the valley on a calm morning. But all in all, the second latest harvest of the 90's ('91 taking that honor) proved to be quite good. We started two weeks late on September 19th dancing with nature's heat, fog and rain. Yields were below normal, which helped us reach our desired maturity levels. Color and acid were excellent. Our typical house style of fresh "zinberry" aromas precedes a fetching brown sugar note. In the mouth there's more impact from acidity than tannin, al Chianti. This is classic Dry Creek Zin with above aging potential similar to '96,'92, or '88.

Cellar Notes: At a recent celebration event, all fifteen Nalle Zinfandels (1984-1998) were poured. To the surprise of many, the older wines were still drinking well. Of course, their fruitiness (which I hold in high regard) was diminished, but complexity and balance kept them charming.

<u>1997 Nalle Zinfandel</u>-Fast track wine. Gorgeous, ripe strawberry nose with Zin spice for interest. Smooth, lush mouth. Tannins seem almost sweet. Likely to peak about 2002.
<u>1996 Nalle Zinfandel</u>-A "small berry" year, this reminds me of '88 and '75 (developing slowly, then drinking well for a decade or more). Starting to assert itself with earthy complexity. Nice entry but tannins still need time to be optimal. Don't rush it.
<u>1995 Nalle Zinfandel</u>-Rounding into its prime. High tone spice and raspberry nose. Higher than usual alcohol and acid but lower tannins give a warm, soft impact to the mouth.
<u>1994 Nalle Zinfandel</u>- A ripe, round vintage and one of the best in twenty years. Complex cedar, jammy nose with superb balance, front to back. Drink or hold.

Nalle Winery also produces 100-200 case lots of Russian River Pinot Noir and barrel fermented Dry Riesling from Mendocino County.

Niebaum Coppola Winery

Niebaum Coppola Winery
1991 St. Helena Hwy
Rutherford, CA 94573
Phone: 707-968-1100 Fax: 707-963-9084
Website: www.niebaum-coppola.com

Niebaum-Coppola was created by Francis and Eleanor Coppola in 1975 when they purchased 1600 acres of the original Inglenook Estate in Rutherford, Napa Valley. The Coppolas began producing wine commercially with the estate's flagship wine, Rubicon, a Bordeaux red blend, with the 1978 vintage. Later, in 1995, the Coppolas were able to buy the rest of the Inglenook Estate, including the Chateau and 90 acres, thus rejoining the historic property.

Zinfandel was the first grape planted on the estate in 1855 and is the only variety that has been continuously planted since inception. The Niebaum Coppola Estate Winery produces a very dense and exotic zinfandel, bottled as "Edizione Pennino Zinfandel." The namesake, Mr. Francesco Pennino, was Francis Coppola's grandfather and the label was taken from Francesco's music publishing company's letterhead.

Tasting room is opened daily from 10-5, with tours at 11 and 2.

Owners: Francis and Eleanor Coppola **Winemaker**: Scott McLeod

1998 Edizione Pennino Zinfandel

Vineyard: Estate grown
There is not much Zinfandel left in Rutherford, long since removed for the more pricey cabernet. But it is the wine that the owners like to drink, and so zinfandel remained. We make Pennino to be saturated and dense, but with the jammy zinfandel fruit and appeal that makes it so unique. All our zinfandel is organically grown and certified, and placed on the exposures that provide great daytime heat and sunlight but that creep into the shadows by 5 o'clock.

Composition: Three 'century clones' of zinfandel, small addition of Cabernet Franc and Cabernet Sauvignon.

Vinification and Maturation: All zinfandel vineyards are picked three times, selecting ripe clusters only. Further sorted, and crushed to three ton open top fermenters. Punch downs by hand. Aged in French and American oak barrels and puncheons.

Related information:
Alcohol: 14.5%   **Residual sugar**: Dry
Brix at harvest: varied 24-27 Brix
Bottling Date: April, 2000
Release Date: March, 2000
Production: 1300 cases
Retail: $40.00

Winemaker's Notes:

The '98 Pennino was the first vintage where we were able to use some of our new blocks that were planted to three grand old clones of zinfandel. The results were so favorable that they made the Pennino blend. The wine has a very round and inviting texture and flavors of the interesting zinfandel spice character that sort of defies description. The Pennino must be engaging, and aromatically we look for very forward sweet aromas with smoky barrel influence. The palate has to have the silky, jammy (nearly viscous) weight to it. Our style is more elegant than some, preferring ripe flavors and textures to over-ripe ones. Above all, Edizione Pennino Zinfandel must be delicious. - Scott McLeod

Cellar Notes:

1988-1998. What is wonderful about the Edizione Pennino Zinfandels is that they are always produced from the same property using similar traditional winemaking techniques. All these vintages reflect the spirit of the zinfandel grape and specifically our estate, and the vintages, their strengths and weaknesses, are part of the wine and apparent.

Orleans Hill

Orleans Hill Viticultural Corporation (aka Orleans Hill)
P.O. Box 1254
Woodland, CA 95778
Phone: 530-661-6538 (but hey, we don't answer)

Located in a WWII Quonset hut just outside the peaceful tomato-growing town of Woodland, up the road a piece from UC Davis, Orleans Hill crafts wines from organically-grown grapes from selected vineyards throughout California. The winemaker and president (and tank cleaner) is Jim Lapsley who directs all of the agricultural continuing education classes from UC Davis and who teaches for the Dept. of Viticulture and Enology. Orleans Hill is his 6000 case "baby" and hands-on labor of love since 1980. Despite teaching wine marketing, Jim does not have a tasting room, mailing list, or wine club. Wines are sold via distributors to select retailers--primarily stores specializing in organic foods.
Owner and Winemaker: Jim Lapsley

Winemaking Philosophy:
We believe in a minimalist approach. Jim's doctorate is in Agricultural History (see his book on the Napa Valley—Bottled Poetry) and he is concerned about the sustainability of farming practices. All of the grapes are organically-grown and Jim uses minimal processing and low or no-sulfite additions to his wines. All of the red wines have no sulfites, which means they are produced to be drunk now—not to be put in a cellar and rot (hey, life is uncertain, right?). The Zin is made from Amador County Clockspring Vineyard grapes and has full flavors—but is made with moderate extraction so that it can be drunk and enjoyed now—not in 5 years!

1998 Clockspring Vineyard, Amador County Zinfandel

Appellation and Vineyard: Well, it is from the Shenandoah Valley, but we put "Amador County" on the label. 100% from the Clockspring Vineyard owned by Frank and Kathy Alviso. Frank's ancestors came into California with DeAnza (the town near Milpitas is named for the family) and as cattle ranchers, the "clockspring" was their cattle brand.

The Clockspring Vineyard is Amador county's first registered organic vineyard. Frank uses cover crops to insure fertility and insect predators for pest control. And the grapes are yummy!

Composition: 100% Zinfandel as far as I know.

Vinification: Standard crushing and cap manipulation for a medium-bodied Zinfandel. Warm fermentation in the 80 degree range, fermented with toasted American oak chips (hey, they all do it—but most won't admit it) and run through a full malolactic. No sulfite added. Filtered and bottled in early 2000 to insure fruitiness.

Maturation: So what is maturity? This wine is soft, supple, has medium tannins, and delivers a mouthful of pleasure. It is not designed to age and it won't benefit from aging. Life is uncertain. Drink dessert first! Actually, we made the wine this way deliberately since we don't add sulfur dioxide. What do you expect from a 100% Amador Zinfandel that retails for about $10.00?

Related information:
Alcohol: 13.35%
Harvest Date: September 26, 1999
Acid: 7g/l
Sulfur Dioxide: 8ppm

Orleans Hill, Amador County

Peachy Canyon Winery

Peachy Canyon Winery
Rt. 1 Box 115 C
Paso Robles, CA 93446
Phone: 805-237-1577 Fax: 805-237-2248

Peachy Canyon Winery and Vineyards were started in 1988 by Doug and Nancy Beckett, both former school teachers. Peachy Canyon farms 80 acres of premium wine grapes and produces 28,000 cases of red wine annually. Doug first started in the wine business in 1982 as a partner in Tobias Vineyards which is no longer in existence.

Tasting Room and picnic area located at 1400 North Bethel Road, Templeton are open daily 11:00 am to 5:00 pm, including major holidays. Wine Club.

Proprietor: Doug and Nancy Beckett   **Winemakers**: Doug Beckett and Tom Westberg

1998 Benito Dusi Ranch

Vineyard Designation: Benito Dusi Ranch is located in the bench of the Salinas River just south of Paso Robles. This area is locally known as the Templeton Gap. A gap or notch in the Santa Lucia Mountains allows a strong flow of marine air to cool this area of the southern Salinas Valley by 10 to 12 degrees on a typical summer afternoon. Zinfandel benefits from this moderation in heat by producing grapes with excellent acid/sugar balance, fully ripe without raisin character. Additionally the vines are dry-farmed, head-trained and average about 75 years of age. Benito takes excellent care of these vines and the end result is consistent and great wine.

Composition: 100% Zinfandel

Vinification and Maturation: Picked at four different dates the grapes were destemmed into 4x4 bins or small tanks. After a 24 hour soak the must was inoculated with D254 yeast plus nutrients. The bins were hand punched 3 times a day while the tanks pumped over 3 times a day. Fermentation was completed in 9 days and the must was gently pressed and combined into two lots. The wine was finished in 59 gallon French and American barrels. After three rackings and 15 months later the wine was lightly filtered and bottled.

Related information:
Alcohol: 14.5%
Brix at Harvest: 25.2
Release Date: 03/09/2000
Production: 715 cases sold at winery and distributed.
Retail: $26.00

W i n e m a k e r ' s   N o t e s : Dark ruby color with black pepper and ripe black cherry fruit aromas. Incredible flavor with balance and intensity in the mouth. Layered fruit integrated with soft and mature tannins on the palate. This wine is an experience you shouldn't miss. A long finish completes the package.

1998 "Estate" Zinfandel

Vineyard: Peachy Canyon Estate, 33%; Mustang Spring Ranch, 66%.
Peachy Canyon vines are about 10 years old, head-trained and dry-farmed. The wood is primarily taken from the Benito Dusi Ranch and Clone 1 Zinfandel grafted to 110-R rootstock. Mustang Springs Zinfandel ranges from 20 to 55 years old. Although the clone is not known, the fruit has small berries and clusters indicative of the older varieties. Both of these vineyards are very low yielding with intense color, flavor and good tannin structure.
Composition: 100% Zinfandel
Vinification and Maturation: The grapes were destemmed into ½ ton and 1 ton bins. Cultured yeast and nutrients were added and the bins were hand-punched 3 times a day until primary fermentation was complete. The must was pressed into small tanks to finish malolactic fermentation and then barreled into small French and American oak barrels. After aging for about 15 months the wine was blended, lightly filtered and bottled.

Related information:
Alcohol: 14.4%
Brix at Harvest: 25.3
Release Date: 03/09/00
Production: 715 cases sold at winery
Retail: $30.00

W i n e m a k e r ' s   N o t e s : Beautiful dark garnet color. The nose is jammy with ripe fruit, strawberry, mint and floral tones. Big in the mouth, complex flavors with fruit, spice and good tannin carrying into a long aftertaste.

Peachy Canyon Winery, Paso Robles

There is a lot of power and concentration in this Estate blend. Will age well and mature in 3-5 years.

1995 Westside Zinfandel

Vineyard: The Westside designation in the Paso Robles appellation refers to grapes grown generally West of the Salinas River in the foothills of the Santa Lucia Mountains. The 1998 blend is composed predominately of grapes from the Templeton Gap, where our tasting room is located. This area is cooler than inland areas with warm mornings and a strong marine influence in the afternoons. The vines typically yield less than 3 tons to the acre with excellent color, full maturity with great acid/sugar balance.

Composition: 100% Zinfandel

Vinification and Maturation: The grapes were destemmed into small fermenters. The must was allowed to soak overnight before adding yeast or nutrients. Tanks were pumped over three times a day to extract color, tannin and flavor from the skins of the grapes. The must was lightly pressed into French and American oak barrels for 14 months of oak aging. The wine was bottled after a light filtration.

Related information:
Alcohol: 14.5%
Brix at Harvest: 25.4
Release Date: 05/01/00
Production: 3621 cases sold at winery and distributed.
Retail: $19.00

Winemaker's Notes: A dark garnet color precedes and incredibly rich, fruity, spicy raspberry nose. The wine is big on the palate yet elegant. The fruit and tannins are well integrated with layers of complex flavors with a clean, long finish. This may be on a par with the Mustang Springs as our best offering of 98 Zinfandel.

1998 Paso Robles Zinfandel

Vineyard: An inclusive blend of grapes from throughout the Paso Robles area reflecting the flavors, aromas and styles of several small microclimates. About 50% of the fruit is from vines east of the Salinas River and the balance of the grapes from Templeton and other vineyards west of the Salinas.

Composition: 100% Zinfandel

Vinification and Maturation: Grapes are hand picked into ½ ton bins at full ripeness. The fruit is destemmed into stainless temperature controlled fermenters and allowed to soak for 24 to 48 hours. Nutrients are then added and special yeast culture is started to provide for a healthy fermentation. The juice is aerated and pumped over the cap three times a day to extract color and tannin from the skins. The wine is pressed off at about 1 Brix and allowed to finish malolactic fermentation before barreling. After 14 months in French and American oak the vineyard lots are blended, lightly filtered and bottled.

Related information:
Alcohol: 15.3%
Brix at Harvest: 25.9
Release Date: 4/19/2000
Production: 2930 cases sold at winery and distributed.
Retail: $17.00

Winemaker's Notes: Medium dark ruby color. A soft scent of strawberries with a floral rose hint in the nose. Flavors of intense cranberry and red cherry fruit on the palate in a firm tannic structure. The finish is long and spicy. A great wine to watch develop with age as it softens and becomes more complex.

1998 Mustang Springs Zinfandel

Vineyard: An older vineyard property of Peachy Canyon located just north and west of Paso Robles. The Zinfandel vines are 30 to 60 years old on very wide spacing with a 3-wire trellis. The fruit has small berries and loose clusters. Irrigation has been sparse with most plants receiving minimal survival quantities of water for several years. The grapes have unique qualities associated with old vine Zinfandel.

Composition: 100% Zinfandel

Vinification and Maturation: Picked at full ripeness with good acidity this fruit was destemmed into ½ and 1 ton fermenters. Nutrients and D254 were added the second day to begin a nine day fermentation. The must was gently pressed a settling tank and taken into barrels after ML was complete. 15 months later the wine was gently filtered and bottled.

Related information:
Alcohol: 14.5%
Brix at Harvest: 25.4

Release Date:3/10/00
Production: 1159 cases sold at winery.
Retail: $26.00

Winemaker's Notes: Medium garnet color. Floral nose with an herbal sage and spice character. Concentrated and refined, this is easily the most complex of the '98s. Medium body and firm tannins with good acidity and fruit make this offering quite a mouthful. A touch of oak sweetens the palate and extends the aftertaste. The balance and tannin should give this wine graceful longevity.

Peachy Canyon also produces Cabernet Sauvignon and Merlot.

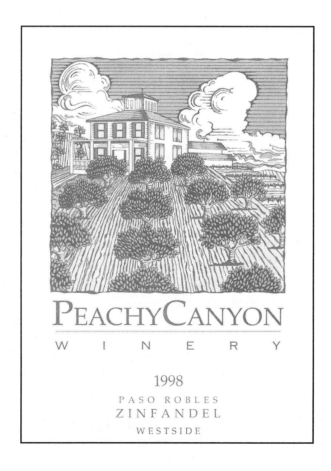

Pedroncelli Winery

Pedroncelli Winery
1220 Canyon Road
Geyserville, CA 95441
Phone: 707-857-3531 Fax: 707-857-3812

Established in 1927, John Pedroncelli Sr. Purchased 90 acres of vineyard along with a winery in the Dry Creek Valley. The Pedroncelli family has been grape and winegrowing for over 70 years in the same spot. They cultivate 105 acres of vineyard and produce 80,000 cases of wine each vintage. Zinfandel was planted on the estate at the time of purchase and continues to be the main variety harvested from the home ranch.

Tasting room open daily from 10 to 4:30 except major holidays. Phone orders, wine club and website all offered. Bocce court open during spring, summer and fall along with picnic area. Local artists exhibit in the attached conference room.

Proprietors: Jim and John Pedroncelli
Winemakers: John Pedroncelli and John Haw

Winemaking Philosophy: Our Zinfandel is made from selected hillside vineyards known to produce a wine exhibiting the intensity, complexity and balance we find to typify this variety. By careful blending, aging and minimal processing, we constantly strive to maintain and improve our wine year after year.

1998 Mother Clone Zinfandel Dry Creek Valley

Vineyard: Pedroncelli, Nivan Buchignani
The Mother Vineyard block, part of the original vineyard purchased in 1927 by John Pedroncelli, is the source of the wine named Mother Clone. It was planted between 1905-1910. Grower Nivan Buchignani also contributes fruit from his west Dry Creek Valley ranch. Budwood from the original vineyard was budded (cloned) onto new rootstock. All vines are head pruned. New planting range in age from 8-25 years old or older. The Petite Sirah that is blended with the Zinfandel comes from Pedroncelli's home ranch. It is adjacent to the Zinfandel vineyards.
Composition: 81% Zinfandel 19% Petite Sirah

Vinification and Maturation: Tank fermented for 6 to 8 days. Pump over with caps sprinkled three times per day for color extraction and cap management 12 months in medium toast American oak barrels (40% new, 60% one to four year old barrels.) Aging potential: 6 years in cellar.

Related information:

Alcohol: 14.4%   **Residual sugar**: 0.20%
Brix at Harvest: 24.4   **Harvest Date**: 9/24 to 9/30, 1998
Bottling Date: January 2000   **Release Date**: May 2000
Production: 13,300 cases sold at winery and distributed.
Retail: $14.00

W i n e m a k e r ' s   N o t e s : Deep purple-red color, full bodied, very plush. Aromas and flavors of ripe blackberry, raspberry, cherry, spice and black pepper. This Zin is balanced harmoniously in flavor, texture and tannin. The rich flavors are generous and lasting.

Food pairing: Chili, Sausages, Pasta with spicy Marinara, Chili rubbed Pork Tenderloin.

1997 Pedroni-Bushnell Vineyard Dry Creek Valley

Vineyard: Pedroni-Bushnell

Located on the eastern bench of Dry Creek Valley, the 20-acre Pedroni-Bushnell Vineyard has been closely associated with the Pedroncellis for 50 years. Winery founder, John Sr., purchased the property then planted it to Zinfandel, Petite Sirah and Carignane in 1945. The vineyard was sold in the 1950's to Al Pedroni, John's son-in-law, who initiated replanting the Zinfandel and Petite Sirah. Al's daughter, Carol Bushnell, inherited the property in 1990 and her husband, Jim, manages the vineyard.

Composition: 95% Zinfandel 5% Petite Sirah

Vinification and Maturation: Fermentation in stainless steel tanks; fermentation took 6 days; aged in American oak barrels (1/3 new oak) for 14 months.

Related information:

Alcohol: 14.46%   **pH**: 3.44
Total acidity: .630

Winemaker's Notes: Intense, deep ruby color. Forward aromas of blackberry, toasty oak and ground pepper. The flavors of wildberries and spice are rich and concentrated. Firm tannins, good acidity and structure point to an extended cellar stay. Aging potential: 10 years in cellar.

Cellar Notes:
1995 Mother Clone Zinfandel: Suggested 8 to 10 years in cellar from release. Doing quite well.
1995 Pedroni-Bushnell Vineyards: Suggested 10-12 years from release
1996 Mother Clone Zinfandel: Suggested 5-8 years in cellar from release. This wine is aging nicely.
1996 Pedroni-Bushnell Vineyard: Suggested 8-10 years from release
1997 Mother Clone Zinfandel: Suggested 10 years in the cellar from release. Rich and luscious, this wine will continue to age well.

Pedroncelli Winery also produces Chardonnay, Fume Blanc, White Zinfandel, Zinfandel Rose` (since 1954), Pinot Noir, Sangiovese, Petite Sirah, Merlot and Cabernet Sauvignon.

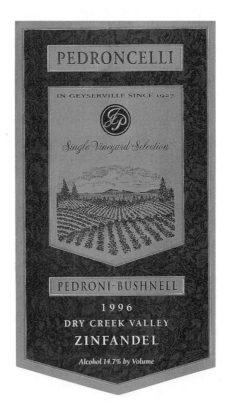

Pedroncelli Winery, Dry Creek Valley

Peterson Winery

Peterson Winery
P.O. Box 1374
1040 Lytton Springs Road
Healdsburg, CA 95448
Phone: 707-431-7568 Fax: 707-431-1112
Email: petersonwinery@juno.com
Tasting and Sales by appointment only.

Peterson Winery is not one of the big corporate wineries with a huge staff churning out hundreds of thousands of cases of wine a year. Peterson Winery makes big wines in small quantities. After 25 years as a winegrower, Fred Peterson truly believes exceptional wines begin in the vineyard. Fred is committed to all facets of grape growing as the foundation for his extraordinary wines. This is why Fred considers himself a winegrower rather than a winemaker.

Owner and Winegrower: Fred Peterson

The Vineyards:

The grapes for Peterson wines come from some of the best vineyards in Northern California. Norton Ranch, a benchland vineyard located in Dry Creek Valley, produces grapes for Peterson Zinfandel, Merlot, Sangiovese and Barbera. On the steep hillside of the Bradford Mountain vineyard on the western edge of Dry Creek Valley, Fred grows Cabernet Sauvignon, Merlot, Petite Verdot, Cabernet Franc and Zinfandel. Peterson Winery produces several wines that are 100% mountain fruit. A percentage of the Bradford Mountain Merlot and Zinfandel are blended with benchland fruit to produce the Peterson Dry Creek Merlot and Dry Creek Zinfandel, adding more intensity and richness to the wine.

At Floodgate Vineyard, located at the far west end of Anderson Valley, the Peterson Winery's Chardonnay and Pinot Noir are grown. Anderson Valley has been compared to the Burgundy region in France, with a very long growing season. Perfect weather to grow exceptional Chardonnay and Pinot Noir.

Winemaking Philosophy:

Only the best of the best will do. Fred carefully selects the grapes that are hand-picked from these vineyards to create the superb hand-crafted Peterson wines.

Fred's winemaking philosophy is zero manipulation. This does not mean the wine is placed in barrels and ignored for months. Zero manipulation means having respect for the grape's quality for that vintage, for each wine to be what the vintage created instead of a preconceived style of wine created through winemaking techniques. Fred allows the natural aromas and flavors of the grape from each vintage to come through, avoiding winemaking methods that might diminish or mask the natural characteristics of the fruit.

1998 Dry Creek Valley Zinfandel

Appellation: Dry Creek Valley
Composition: 84% Zinfandel, 12% Petite Sirah, 2% Carignane, and 2% Mourvèdre.

Related information:

Alcohol: 13.8%	**Residual sugar**: Dry
Brix at Harvest:25	**Harvest Date**: 9/25/98
Release Date: July 2000	**Production**: 1000 cases
Retail: $17.00	

Winemaker's Notes:
Here at Peterson Winery we are quite pleased with the 1998 version of our Dry Creek Valley Zinfandel. The vintage is really a stylistic combination of the 1996 and 1997. The 1996 was rather light and tight, a nice wine with high natural acidity but tons of fruit. The 1998 Zinfandel is well balanced. The wine was aged for 10 months in older French oak barrels and bottled unfined and unfiltered.

Peterson Winery also produces Sangiovese, Merlot, Chardonnay and Cabernet Sauvignon.

Pezzi King Vineyards

Pezzi King
3225 West Dry Creek Road
Healdsburg, CA 95448
Phone: 707-433-8785 or 800-411-4PKV Fax: 707-433-6838

    Pezzi King Vineyards is family owned and dedicated to producing the highest quality wines possible. Current production is approximately 20,000 cases with growth projected towards additional vineyard designated reds.

Tasting Room and gardens open daily from 10:00 am to 4:30 pm. Mail order. Wine Club. Special Events.

Proprietors: James P. Rowe and Family   **Winemaker**: Cecile Lemerle
President: Jim Rowe Jr.
Assistant Winemaker: Tim Crowe
V.P. Sales and Marketing: Tom Rozner
Director of Marketing: Ian MacNeil

Winemaking Philosophy: Pezzi King's philosophy is to produce intensely flavored and concentrated wines that achieve a balance between current drink-ability (supple tannins, well integrated vivid fruit tones and appropriate acidity) and aging potential. Quality not quantity will always be our ultimate objective.

1998 Estate, Dry Creek Valley Zinfandel

Vineyard: Pezzi King's Estate Zinfandel is grown in terraced hillside vineyards 600 to 900 feet above Dry Creek Valley. Low yields, volcanically derived soils and judicious use of the finest French oak barrels contribute to the PKV Estate Zinfandel.
Composition: 100% Zinfandel
Vinification and Maturation: PKV Zinfandel is vinified in a mix of open and closed top stainless steel fermenters at 80-85°F. A combination of pumping over and punching down is used to achieve maximum color and phenolic extraction. 75-80% whole berry fermentation is utilized to elevate the intensity of berry aromas and flavors. Aged for 12-15 months

in 1/3 new French oak, 60 gallon barriques with the balance in one to three year old French oak barrels.

Related information:

Alcohol: 14.5%	**Residual sugar**: NA
Brix at Harvest: 24.5	**Harvest Date**: October 19 and 20th, 1998
Bottling Date: April 2000	**Release Date**: July 2000
Production: 906 cases	
Retail: $28.00	

W i n e m a k e r ' s N o t e s : Cinnamon, rose petal and blackberry aromas complement flavors of sweet cocoa, vanillin, black pepper and black cherries in this massively concentrated estate PKV Zinfandel.

1998 Maple Vineyard, Dry Creek Valley

Vineyard: Pezzi King's Maple Vineyard Zinfandel is grown in benchland vineyards sitting up above the Dry Creek Valley floor. Low yields, excellent vineyard management and judicious use of the finest French oak barrels contribute to a strong following for the PKV Maple Vineyard Zinfandel.
Composition: 100% Zinfandel
Vinification and Maturation: Process remains consistent for all PKV Zinfandels.

Related information:

Alcohol: 14.0%	**Residual sugar**: NA
Brix at Harvest: 24.1	**Harvest Date**: October 1-12, 1998
Bottling Date: April 2000	**Release Date**: July 2000
Production: 500 cases	
Retail: $27.00	

W i n e m a k e r ' s N o t e s : Raspberry, blackberry aromas and flavors are completed by clove, cherry and cigar box notes in the finish of the 1998 PKV Maple Vineyard Zinfandel.

1998 Old Vines, Dry Creek Valley

Vineyard: Pezzi King's inaugural vintage of "Old Vines" Zinfandel makes its debut with the 1998 vintage. The "Old Vines: Zinfandel is derived from several old vine vineyards located in various parts of Dry

Creek Valley. Low yield, dry farming, old vine age and judicious use of the finest French oak barrels suggest the PKV 1998 "Old Vine Zinfandel

will be among the finest Zinfandels in California.
Composition: 100% Zinfandel
Vinification and Maturation: Process remains consistent for all PKV Zinfandels.

Related information:
Alcohol: 14.5%
Brix at Harvest: 24.6
Harvest Date: October 7th - November 11th, 1998
Bottling Date: April 2000 **Release Date**: July 2000
Production: 2800 cases   **Retail**: $27.00

W i n e m a k e r ' s   N o t e s :   Raspberry, licorice and apricot aromas are complemented by raspberry, clove and nutmeg flavors. Classic old vine Zin complexity.

Pezzi King also produces Cabernet Sauvignon, Fume Blanc, and Chardonnay. Merlot, Syrah and Late Harvest Sauvignon Blanc.

Phoenix Vineyards and Winery

Phoenix Vineyards and Winery
3175 Dry Creek Road
Napa, CA 94558-9722
Phone and Fax: 707-255-1971
Email: phoenix@fcs.net Website: www.phoenixvineyards.com
Proprietors: The Bader Family
Winemaker: Aaron Bader **Vineyard Manager**: David Bader

Winemaking Philosophy:

Like most of the wines that we make, our Zinfandel program starts in the vineyard. We feel the best wines are grown that way. Quality clonal material, used in conjunction with great vineyard sites, gives superior quality grapes. Also, we feel that optimal ripeness is the key to our ideal Zinfandel.

We try our best to let the wine make itself. We use gentle techniques such as punching-down the cap (instead of pumping over) and basket pressing to avoid mechanical damage to the skins, stems, and seeds. We also like to employ 20 to 40% stem retention at the crusher for more extract and black pepper notes in the wine. To help showcase the pure fruit essence of the varietal we use only a portion of new barrels for our Zinfandel. We do not fine or filter our red wines.

Current Releases:

On the basis of quality judgments we did not produce a 1998 vintage Zinfandel. We have a 1999 in barrels currently. This wine should be ready for bottling in summer 2000 and released in the summer of 2001.

1997 Napa Valley Zinfandel

This wine is showing very well right now. The fruit and oak are in superb balance. The fruit characters of raspberry, and black cherry are quite ripe and rich in this wine. The wine is quite lively and bright despite the 14.6% alcohol level. If you are lucky enough to have some, you will find out that it is going to go very well with everything that comes off of your backyard grill for the next few years.

1996 Napa Valley Zinfandel

The wine has a little more structure than the 1997. For this reason it is in a brooding phase currently. With exposure to air for 20 minutes the fruit characters start to come forward. This vintage has more jammy fruit and some cassis notes when compared to the 1997. It would do very well with grilled foods, stews or Osso Bucco.

Phoenix Vineyards and Winery also produces Cabernet Sauvignon, Sangiovese (Blood of Jupiter), Merlot (Little Blackbird) and Pinotage.

Quivira Vineyards

Quivira Vineyards
4900 West Dry Creek
Healdsburg, CA 95448
Phone: 707-431-8333, or 800-292-8339 Fax: 707-431-1664

Located on the west side of Dry Creek Valley in Northern Sonoma County, the Quivira Vineyards' ninety acre estate is the primary source of select grapes used to produce Quivira wines. The land was purchased in 1981 and the winemaking facility built in 1987 with a capacity of 20,000 case production per year. Zinfandel already existed in the vineyards of Quivira, having been planted in the early 1960's and has been produced by the winery since 1983.

Tasting Room and picnic area open daily from 11:00 to 5:00. Mail order. Wine Club. Tours by appointment.

Proprietor: Holly and Henry Wendt   **Winemaker**: Grady Wann, 1990

Winemaking Philosophy: Quivira Zinfandel is made in a style that highlights the ripe raspberry and blackberry fruit character from our hillside vineyards. Picked ripe and hand sorted, the grapes are gently processed to capture the ripe fruit flavors without excessive tannins. The soft tannin structure is complimented by aging in mostly French oak cooperage for ten months with frequent racking. Petite Sirah is blended to add complexity and balance by contributing depth of color, tannin, as well as its pepper and black cherry character to the Zinfandel blend.

1998 Zinfandel Dry Creek Valley

Vineyard: Quivira Zinfandel is a blend of four different vineyards all located in Dry Creek Valley. Forty percent of the blend is from Quivira's estate vineyards where the hillsides are all planted to Zinfandel. The cornerstone for Quivira Zinfandel is the Zin planted in 1962. This vineyard has been represented in every Quivira Zinfandel since it's first vintage in 1983. The remaining hillsides of the estate property were planted to Zin in 1991. In the last few years the Zinfandel that comes from the valley floor, also planted in 1962, has been of high enough quality to add to the blend. The three vineyards Quivira buys fruit from are as follows: The Dieden Vineyard located on the eastern benchlands of Dry Creek Valley with vines averaging twenty years of age. The Dieden Vineyard sees the late afternoon sun which provides for ripe Zinfandel with lots of black fruit character. The Tambollini Vineyard is located on the western slopes of Dry Creek Valley and arranged in three different blocks. The first block was planted in the early 1900's, the second planted in the early 1970's, and the last block was planted in 1989. Our third source of Zinfandel comes from the Standley Ranch, also located on the western slopes, is a single acre of Zinfandel planted in the early 1900's.

Composition: 13% Petite Sirah 87% Zinfandel

Vinification and Maturation: Hand picked and hand sorted at the crusher. Fermented in open top fermenters eight to ten days. Gentle pump over cap management. Malolactic fermentation takes place in the barrel. Four rackings throughout the year with some oxidation induced in the early rackings to help soften tannin and make the wine more aromatic. Light egg white fining and filtered before bottling. Aged ten and a half months in new to three year old French oak barrels, medium plus toast. Ten months bottle age before release.

Related information:

Alcohol: 13.5%   **Residual sugar**: Less than 0.1%
Brix at Harvest: 24.5   **Harvest Date**: 10/08 - 10/28
Bottling Date: September 99   **Release Date**: May 1st, 2000
Production: 6,900 cases sold at winery and distributed.
Retail: $18.00

Winemaker's Notes: Ripe berry fruit and a bit of oak compliment the nose. Nice fruit and oak balance with great texture on the palate. Luscious blackberry fruit entry with a little oak in the mid-palate and in the finish.

Quivira Vineyards also produces Sauvignon Blanc and Dry Creek Cuvée (a Rhone style blend)

Rabbit Ridge

Rabbit Ridge
3291 Westside Road
Healdsburg, CA 95448
Phone: 707-431-7128

Rabbit Ridge is located in the Russian River Valley of Sonoma County and buys Zinfandel grapes come from all over the entire state of California. Currently we produce ten different Zinfandel wines.

Tasting room

Proprietors: Erich Russell   **Winemaker**: Erich Russell and Susie Selby

Winemaking Philosophy:

All of our Zinfandel lots are individually fermented as separate lots. The wines receive a maximum amount of pumpovers for maximum skin contact - then spend anywhere from three to six weeks of extended skin contact time. This produces bold, spicy, and jammy Zinfandels.

1998 California Barrel Cuvee Zinfandel

This is our entry level Zinfandel made from grapes from Amador, Mendocino, San Luis Obispo, and Lodi. It is medium bodied with raspberry and cherry spice characteristic of Rabbit Ridge Zinfandels.

Related information:
 Production: 36,000 cases
Retail: $9.00

1998 Sonoma County Zinfandel

This is the flagship wine of the winery. It is produced from a variety of vineyards in the best growing regions of Sonoma County. It is a full bodied, very extracted blend of flavors highlighting blackberry and boysenberry. The 1998 vintage was very low yielding so our production went from 33,000 cases of the 1997 to 12,000 cases of the 1998.

Related information: **Production**: 12,000 cases **Retail**: $15.00

1998 Olson Vineyard Zinfandel

Very small production single vineyard wine. The grapes come from a small, hillside vineyard in Dry Creek Valley. This wine is the essence of blackberry jam.

Related information:
Retail: $25.00

1998 Hedin Vineyard Zinfandel

This Russian River vineyard produces wines that are the essence of raspberry. At this time in March, 2000 this wine is still in barrels - which is very unusual for us. However, the wine is so big and high in alcohol that it needed an extra six months in barrels. The wine will be bottled in the summer of 2000.

Related information:
Production: 1500 cases
Retail: $25.00

1998 Winemaker Grand Reserve Zinfandel

This wine is the best barrels of all the best lots of Zinfandel in the winery. The production is always limited and it is the highest example of Rabbit Ridge Zinfandel. This wine will not be bottle until the summer of 2000 and there will be less than 1000 cases.

Related information:
Production: 1000 cases
Retail: $35.00

Ravenswood

Ravenswood
18701 Gehricke Road
Sonoma, CA 95476
Phone: 707-938-1960 or 1-888-669-4679 Fax: 707-938-9459
Website: www.ravenswood-wine.com

Ravenswood is a specialty winery in Sonoma, created in 1976 by Joel Peterson and his business partner, W. Reed Foster. Two thirds of the production is Zinfandel, with the balance split between Merlot, Cabernet Sauvignon and a very small amounts of barrel-fermented Chardonnay. Each wine produced at Ravenswood has a distinct vintage and vineyard character as well as style that is uniquely Ravenswood. The Zinfandels, with the exception of the Vintners Blend, are from grapes grown in the Sonoma and Napa Valleys and in some cases are vineyard designated.

Tasting Room and picnic area open daily from 10:00 to 4:30. Mail order. Wine Club. Tours by appointment. Summer week-end BBQs.

President and Winemaker: Joel Peterson
Assistant Winemaker: Peter Mathis **Enologist**: Benedict Rhyne
CEO and Chairman: W. Reed Foster

Winemaking Philosophy:

While the Vintners Blend series (Zinfandel, Merlot and Chardonnay) stresses early accessibility, Ravenswood's other hand-crafted releases posses a unique vintage and vineyard identity in addition to an intense, powerful character with which the winery has come to be associated. These attributes arise from what can only be called stubborn and impractical Old World enological practices. Our wines ferment in small wooded tanks and rely upon wild, natural yeasts. Grape stems are frequently added for structure, and the cap of skins is punched by hand three to five times per day. We age the wine in small French oak barrels, fine it gently with egg whites if needed, and eschew filters and centrifuges as much as possible because we believe they tend to diminish flavor.

Many of our carefully selected grapes come from dry-farmed 70 to 100 year old vineyards which yield low crops of highly concentrated fruit, and our younger vines are meticulously farmed and regularly thinned to ensure premium quality. Our wines from vineyards that consistently exhibit distinct flavor and excellence are vinified and aged separately with the vineyards designated on the bottles.

The resulting wines - rich, complex, fully varietal and true to location are well suited to aging. They will generally reach their peak in seven to ten years, although most are drunk young and greedily by people who love big, gutsy, unapologetic wines.

Ravenswood has a number of Zinfandels from specific counties and from specific vineyards. We have learned over the years that each area has a unique flavor and character that it brings to the wine. We at Ravenswood enjoy this diversity of flavor as part of the pure, multifaceted expression of wine. It is our hope that the intelligent wine consumer will also derive mental stimulation and organoleptic pleasure from this diversity.

In 1990 the winery adopted the slogan (most likely against its will) "NO WIMPY WINES" and has since developed many variations on that theme. The winery's philosophy is that they take their winemaking and distribution seriously, indeed, but try to have plenty of fun along the way and not take themselves too seriously.

The personalities connected with Ravenswood vary greatly, but they are all dedicated people who love forthright, unblushing wines and believe the age-old art of winemaking is an unceasing adventure and reward.

1998 Sonoma County Zinfandel Old Vine

Vineyard: Sonoma County Zinfandel tastes the way great Zinfandel is supposed to taste; full, spicy and richly berry-like, but balanced with firm astringency. Our blend, predominately from several vineyards in the northern part of the County, has the black cherry and mint characteristics that we've come to expect from that region. Many of the grapes are from 60 to 80 year old vines on the Dry Creek bench land west of Healdsburg - a location that many wine enthusiast (including, presumably, the intuitive Italians who originally chose to plant in this area) consider ideal for Zinfandel. Another substantial portion comes from old hillside vines in Sonoma Valley that add an intense spicy character and ample body. This wine carries the rich, opulent and full-flavored character that has become the Ravenswood trademark.

Composition: 91% Zinfandel and 9% mixed blacks

Vinification and Maturation: Our wines ferment in small wooded tanks and rely upon wild, natural yeasts. Grape stems are frequently added for structure, and the cap of skins is punched by hand three to five times per day. We age the wine in small French oak barrels for approximately 16 months, fine it gently with egg whites, and eschew filters and centrifuges as much as possible because we believe they tend to diminish flavor.

Related information:
Alcohol: 14.5%
pH: 3.52
TA: 6.72
Production: 7982 cases sold at winery and distributed.
Retail: $15.25

W i n e m a k e r ' s   N o t e s :
An attractive ruby color and aromas of dark fruit, licorice and spice are your entr'acte to the classic Zinfandel flavors of black cherry, blackberry, dark plum, caramel and pepper. Zesty fruit acidity not withstanding, the texture in the mouth is pleasantly plump, with tannins that are fine grained and integrated. As is so often the case with the rich reds of Ravenswood, there is a fine balance to the wine, which offers attractive drinking at an early age in addition to the substance and structure for aging. Five to ten years in the bottle would not be at all unreasonable.

1 9 9 8   L o d i   Z i n f a n d e l

Our 1998 Lodi Zinfandel is sourced from several specific vineyards, with a majority harvested from Perrin, Ferreira and Kramer Vineyards, in addition to tiny lots from several sites, including Kirschenman and Spenker. Spicy cherry liqueur aromas introduce the bright, lively flavors of berry, tart pie cherries and dried plum finishing with clove, nutmeg, toasted almond and pepper, supported by moderate tannins and crisp, mouth watering acidity.

While capable of at least 3-5 additional years of bottle age, in this election year one might adapt the Tammany Hall philosophy for voting as a Lodi Zin philosophy for drinking: do it early and do it often!

Vinification and Maturation: Remains virtually the same with all Ravenswood Zinfandel.

Related information:
Alcohol: 14.10%
pH: 3.52

TA: 6.42
Production: 10777 cases sold at winery and distributed.
Retail: $13.75

1998 Mendocino Zinfandel

From the county of Mendocino, just north of Sonoma County, comes this new addition to the Ravenswood family of Zinfandels. 100% Zinfandel, drawn from two distinct vineyards - Bartolomei (89%) and Lalanne (11%) - this youthfully delicious wine sports alluring aromas of dark fruit and smoky oak. Bright and lively at this early stage of life, the texture is round and full and the opening flavors of wild berries, black cherries, plum, pomegranate and dark chocolate are accented with hints of baking spice, black pepper and creamy oak.

While the pleasure of drinking this wine in it's youth is obvious, the advantage of choosing to age this wine an additional 4-6 years would be the evolution of the more subtle, complex flavors of the mature wine.

Vinification and Maturation: Remains virtually the same with all Ravenswood Zinfandel.

Related information:
Alcohol: 15.2%
pH: 3.58
TA: 9.11
Production: 897 cases sold at winery and distributed.
Retail: $13.75

1998 Amador County Zinfandel

From the Sierra Foothills AVA, east of the city of Sacramento in what is often referred to simply as "Gold Country", comes our 1998 Amador Zinfandel. 100% Zinfandel, grown on 35-40 year old controlled yield vineyards that typically produce 3-3.5 tons per acre, this wine is a worthy example of the unique terroir of the Amador region. There are those that have visited the region in the summer who swear that the aromas of the rich, red, sun baked earth are identifiable in the aromas of the wine, along with the bright cherry and wild berry-like fruit and cardamom spice. 12 to 14 moths maturation in new and seasoned French oak has rounded, fattened, and deepened the fruit and cracked-pepper spice flavors while contributing additional notes of vanilla and smoke.

With lively acidity and fine tannins, the Amador is delicious in it's youth, but the structure for additional bottle age is apparent as well. 6-8 years would certainly not be unreasonable.

Vinification and Maturation: Remains virtually the same with all Ravenswood Zinfandel.

Related information:
Alcohol: 14.1%
Retail: $13.75

1998 Napa Valley Zinfandel

The grapes for the Napa Valley Zinfandel are selected from a mix of old vine vineyards, mature and younger vineyards. As in the past, the majority of grapes are harvested from the Perez and Czapleski vineyards, with additional fruit coming from Ballentine, Luvisi, Plam and Wooden Valley vineyards. Our 1998 Napa Valley Zinfandel has bright fresh cranberry and pomegranate aromas, with ripe raspberry and hints of black cherry. A pretty plum color, the wine has lots of bright acidity, carrying the flavors into a long, lovely finish. Sixteen months in French oak rounds out the mouth feel, integrating notes of vanilla and smoke with the lively berry flavors. This year a touch of Petite Sirah (less that 2%) was added, giving the wine just a hint, just a whisper of white pepper and tar. The 1998 Napa Valley Zinfandel is more than approachable now…it is delicious. Cellaring is not out of the question though, with 6-8 years being more than reasonable. Production was down this year, so covet your cases!

Vinification and Maturation: Remains virtually the same with all Ravenswood Zinfandel.

Related information:
Alcohol:14.1
Retail: $14.00

1998 Dickerson Vineyard, Napa Valley

Vineyard: Dickerson Vineyards.
Produced continuously since 1982, cedar, eucalyptus and raspberry scents as distinctive as fingerprints characterize the Zinfandel from Dickerson Vineyards. Physician-cum-wine-enthusiast Bill Dickerson is the guardian of the vines that bestow these superlative wines. His

vineyard on (what else?) Zinfandel Lane in the west side of Napa Valley is sheltered by the Mayacamas Mountains, allowing the long Napa growing season to build up full, round flavors in the grapes. The confer claret-style Zinfandel at its best - sumptuous, deeply aromatic wine with pungent flavors and a long lasting finish. Crop levels are low, about 1½-2½ per acre, and the vineyard is small, so production is quite limited.
Composition: 100% Zinfandel
Vinification and Maturation: Remains virtually the same with all Ravenswood Zinfandel.

Related information:
Alcohol: 14.10%
pH: 3.53
TA: 6.93
Production: 1707 cases sold at winery and distributed.
Retail: $28.50

W i n e m a k e r ' s   N o t e s :
Bright aromas of raspberry, cherries, cedar and menthol. Juicy, minty, refined sophisticated spice flavors. Substantial tannins with long minty, raspberry vanilla finish.

1998 Cooke Vineyard

Vineyard: Cooke Vineyard, Sonoma County.
Carved out of the woods with a steep exposure in the high eastern most hills of Sonoma Valley, is the eighteen acre Cooke Vineyard. The stony soil of this small vineyard produces a shy crop in numbers (only a half ton per acre) but an abundantly and richly flavored Zinfandel. After working with this vineyard for several years, we decided to elevate the Cooke wine to one of our traditional vineyard-designated Zins, joining Old Hill Ranch and Dickerson Vineyards. We think the singular character of these grapes, matched with Ravenswood's traditional winemaking practices, will provide many memorable Zinfandel experiences in the years to come. The Cooke Zinfandel is a claret-style wine with lots of fruit flavor and tannins, but even in its youth the wine is balanced with fresh black cherry and vanilla flavors dominating.
Composition: 100% Zinfandel
Vinification and Maturation: Remains virtually the same with all Ravenswood Zinfandel.

Related information:
Alcohol: 15.3%
pH: 3.82
TA: 8.04
Production: 174 cases
Retail: $31.00

Winemaker's Notes:
Very dense, blackberries, red cherries, pepper and mind. Very intense, penetrating flavor. Lots of tannin balanced by a long lush sustaining finish. Perhaps the biggest and most long lived of the 98 single vineyard Zins.

1998 Belloni Vineyard

Vineyard Designation: Belloni, Russian River
Situated in the cool, often foggy, Russian River Valley near Fulton, this 88 year old vineyard is one of our last Zinfandel vineyards to be harvested. Originally planted in the early 1900's, I started working with the Belloni Zinfandel in 1990. The wine was so good that Belloni made vineyard-designated status the very next year. Due to the long hang time Belloni develops intense fruit flavors that are likened to boysenberry and plums. This might be rather heavy but for the acidity maintained by the cool regions characteristics that keep the wine beautifully balanced.
Composition: 100% Zinfandel
Vinification and Maturation: Remains virtually the same with all Ravenswood Zinfandel.

Related information:
Alcohol: 13.70%
pH: 3.48
TA: 7.77
Production: 549 cases
Retail: $28.50

Winemaker's Notes:
Complex aromas of blackberry, boysenberry and licorice highlighted by tones of cracked black pepper. Bright with very fresh notes of spice make this wine a joy to consume with rich, herbal foods.

1998 Monte Rosso

Vineyard: Monte Rosso, Sonoma Valley
This wonderful old vineyard on Moon Mountain has been owned and meticulously cared for by the Martini family of Louis Martini Winery for many years. The turn of the century vines that produced these Zinfandel grapes are picture perfect with their long, graceful, head pruned arms and their hillside setting overlooking the Sonoma Valley.

Traditional folklore in the Sonoma Valley dictates that planting red grapes in the red volcanic soils of these mountain will produce terrific red wine. The wine from this vineyard is a testimony to that wisdom. Deep, rich, dark color hints of the concentration of this wine. The nose of cherries, plums, leather, pepper and spice are a perfect introduction to the supple, round, full bodied spicy flavors of Monte Rosso Zinfandel.
Composition: 100% Zinfandel
Vinification and Maturation: Remains virtually the same with all Ravenswood Zinfandel.

Related information:
Alcohol: 15.2%
pH: 3.72
TA: 7.48
Production: 1228 cases
Retail: $28.50

Winemaker's Notes:
Bright distinct aromas of raspberry, cranberry, cedar, camphor and mind. Ripe, juicy flavors are nicely balanced with tannins and vanilla oak highlights. The wine is complex and refined. It is a wine that even people who claim not to like red wine will often enjoy.

1998 Old Hill Ranch

Vineyard: Old Hill Ranch.
This unique vineyard in Sonoma Valley is an extension of the Oak Hill Organic Farm, saved from oblivion by Otto Teller, a sage octogenarian who employs no pesticides or irrigation. The 110 year old vines are situated on a gravely, well drained hill that receives a full share of summer sun as well as cooling breezes. The crop level is seldom higher than one and a half tons per acre, and this scant tonnage - combined with the ideal growing conditions - produces exceptionally intense, spicy fruit. Old Hill Zinfandel is dark and rich with provocative flavors of

black-berries, mint and black pepper. Its unparalleled depth and complexity will reward those who have the patience to put a few bottles away in the cellar.

Composition: 100% Zinfandel

Vinification and Maturation: Remains virtually the same with all Ravenswood Zinfandel.

Related information:
Alcohol: 14.80%
pH: 3.74
TA: 7.40
Production: 447 cases
Retail: $33.00

Winemaker's Notes:
Concentrated blackberry, violet, black pepper, dark chocolate and mint remind the taster of wine from the northern Rhône Tight and a little closed (at this writing) as is typical of Old Hill in it's youth, the wine exhibits dark berry flavors and substantial tannins with a long powerful finish.

1998 Kunde

Vineyard: Kunde, Sonoma County
Each year Ravenswood gets a small portion of Kunde's Century old Shaw Ranch Vineyard. Wines from this vineyard are lush and concentrated. The climate in the Kenwood area along with the deep, well drained volcanic soils, conspire to produce grapes with soft ripe, well deserved tannins.

Composition: 100% Zinfandel

Vinification and Maturation: Remains virtually the same with all Ravenswood Zinfandel.

Related information:
Alcohol: 15.10%
pH: 3.45
TA: 8.15
Production: 189 cases
Retail: $28.50

Winemaker's Notes:

Jammy, ripe, sweet fruit aromas of black cherries and plums combined with scents of chocolate, coffee and pepper announce the Sonoma Valley origins of this wine. Intense and round with substantial tannins on the palate this wine should develop nicely.

1998 Big River

Vineyard: Big River, Alexander Valley (formerly known as Black Mountain)
A wonderful low yield, 80 plus year old vineyard that was recently purchased by Scott and Lynn Adams. The fruit is remarkably concentrated, but the hillside location tends to thicken the skins which requires a fair amount of tannin management during fermentation.
Composition: 100% Zinfandel
Vinification and Maturation: Remains virtually the same with all Ravenswood Zinfandel.

Related information:
Alcohol: 114.70%
pH: 3.56
TA: 6.51
Production: 688 cases
Retail: $28.50

Winemaker's Notes:
Sappy, pungent aromas of black raspberry, mocha, pepper and roast coffee. Smooth, silky wine that is both concentrated and refined.

1998 Teldeschi Vineyard

What do you get when you put three generations of a hard working Italian family in the perfect grape growing conditions of Dry Creek Valley in northern Sonoma County? The answer? Quite simply, some of the best old vine Zinfandel grown in California. Combine these grapes with great winemaking and you get superb red wine that is just about as good as it gets.

These older (and some younger) low production vines that make up the Teldeschi wine from Ravenswood are a mix of Zinfandel, Carignane and Petite Sirah. This melange is typical of the older plantings in California that use the different nature of each grape to elicit the best wine from the specific conditions of a particular growing region, while also allowing for flexibility in a vintage. The majority grape, Zinfandel, provides a thick ripe cherry berry character, while the Carignane proves bright raspberry flavors and balancing acidity. The Petite Sirah provides deep color, pepper spice and firm tannins. Small open-top fermenters, the use of native yeast fermentation, manual punching down of the cap, minimal processing, and aging in small French oak cooperage all contribute to a style that brings out the best in the vineyard. The three varieties are fermented separately and blended to taste to produce the best wine possible from the vineyard each year.

Related information:
Alcohol: 14.40%
pH: 3.54
TA: 6.41
Production: 2296 cases.
Retail: $28.50

1998 Barricia Zinfandel

Vineyard: Barricia is a terrific old vineyard in one of the best growing locations in Sonoma Valley. Well drained volcanic soils keep vine vigor low and crop production moderate. Low yield combined with intense sun exposure produce wines of dramatic intensity and complex flavor.

Patricia Heron (the first woman Superior Court judge in Contra Costa) and her partner Barbara Olsen contracted their names to form the vineyard name Barricia, a word that coincidentally means barrel in Portuguese. Pat, a long time wine lover, has worked hard to produce grapes that would be consistent of a quality worthy of a vineyard designation. She has succeeded admirably. The wine has spicy vibrant aromas of vanilla, blackberry, and cracked peppercorn. It is sweet and lively in the mouth with excellent density to match the substantial ripe tannins and the slight note of coffee and dark chocolate in the finish.

1998 Vintners Blend Zinfandel

Vineyard: California
Composition: 100% Zinfandel

Vinification and Maturation: Remains virtually the same with all Ravenswood Zinfandel.

W i n e m a k e r ' s   N o t e s :
Each year I comb the countryside to seek out wineries with small lots of well-made Zinfandel. These wines, blended with up to roughly 60% of Ravenswood's own crushed and fermented Zinfandel, produce a style that is distinctly Ravenswood. Careful choosing and blending achieve a well structured wine with charming, youthful, Zinfandel fruit character.

The Vintners Blend has a rich, somewhat soft, moderately complex, spicy, ripe, raspberry aroma. The flavors are those of black berries, mint and vanilla with a sturdy, slightly astringent finish. The freshness and youth that we forge in the Vintners Blend allows it to be released with slightly less barrel and bottle age. We pass our economic savings on to the consumer who will find these forceful but friendly wines to a wonderful complement to pasta, poultry, red meats and other highly flavored dishes.

Ravenswood also produces Merlot, Petite Sirah, Cabernet Franc, Cabernet Sauvignon, Chardonnay, Dry Gewürztraminer, Late Harvest Gewürztraminer, Icon (Rhône-style) and several vineyard designate Bordeaux-style wines.

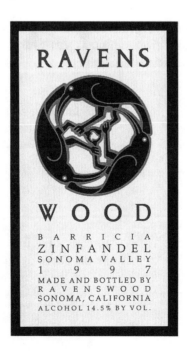

Renaissance Vineyard and Winery

Renaissance Vineyard and Winery
P.O. Box 1000
Oregon House, CA 95962
Phone: 800-655-3277 Fax: 530-692-2497
Website: http://www.renaissancewinery.com
Tasting room open Friday and Saturday by appointment only.

To me, Renaissance is red wine. Its volcanic iron-rich soil, its well exposed, steep, granite slopes, the seemingly endless diversity of microclimates in our estate vineyard, all enable us to produce a large variety of reds normally associated with wine regions separated by great distances and climate conditions. **Winemaker**: Gideon Beinstock

1998 Zinfandel

Appellation: Sierra Foothills
Composition: 100% Zinfandel
Vinification and Maturation: Hot fermentation in stainless steel tanks, aeration and sprinkle-over cap management for 14 days. Aged in neutral oak barrels and ovals for 13 months.

Related information:
 Alcohol: 13.1%   **Residual sugar**: Dry
pH and TA: 3.6 / 6.6   **Brix at Harvest**:23.6
Harvest Date: 10/02/98   **Bottling Date**: 12/01/99
Release Date: 4/01/2000
Production: 365
Retail: $17.00

Winemaker's Notes:
Thick and dense. Complex. Tidal. Aromatic and fresh. A layer of ripe fruit. Slippery, slick texture. Oily. Aromatic in the sense of the Muscat grape. Rolling, voluptuous. We compare it to a Monet where the colors represent the complexity of the palate. There is an orchestra playing right now in your mouth. Jazzy, round, sweet and easy.

Renwood Winery

Renwood Winery Inc.
12225 Steiner Road
Plymouth, CA. 95669
Phone: 209-245-6979 Fax: 209-245-3732
www.renwood.com
info@renwood.com

Surrounded by some of America's oldest Zinfandel vineyards, Renwood Winery is located in the picturesque California foothills of the Sierra Nevada mountains. Renwood was founded in 1992 by a group of investors led by Boston native Robert Smerling. Mr. Smerling fell in love with Amador County with its rich history and magnificent vineyards. A journey was begun to produce wines from the finest vineyards in Amador County. Wine production has grown from 2,500 cases in 1992 to over 30,000 cases today.

Tasting Room open daily from 10:30 to 4:30.

Proprietor: Robert Smerling, chairmen and founder. **Winemaker**: Don Reha

Winemaking Philosophy: Extending maceration and controlling extract create wines that capture the flavor of their distinct vineyard sites without using other blending grapes. Only the finest oak is used and selected for each vineyard.

1997 "Jack Rabbit Flat " Zinfandel

The 1997 vintage is a clear example of why our "Jack Rabbit Flat" Zinfandel is one of the most sought after wines in the Renwood family of Zins.

The fruit for this wine comes from the Fox Creek Vineyard, which is one of the true old vineyards of Amador County that was planted over 80 years ago. The age of the vines combined with the steeper elevation of the vineyard provided us with small quantities (less than three tons per acre) of rich, concentrated fruit. This immaculately maintained vineyard gives the head-trained vines the ability to ripen to maximum potential. Thus, the fruit is picked at the highest degree of ripeness, resulting in a wine of great richness and complexity.

Related information:
Alcohol: 15.8%   **pH**: 3.37
TA: 0.76 g/100 ml   **Bottling date**: July 1999
Production: 1550 cases   **Retail**: $28.00

Winemaker's Notes: The deep, almost black, color of this wine gives you the first hint of intensity that you will encounter on the palate. This is a bold, full-bodied Zinfandel with aromas of cinnamon stick and exotic dried fruit and flavors of blackberry, cocoa and coffee that linger through the finish. Enjoy the 1997 Renwood "Jack Rabbit Flat" Zinfandel for many years to come - it has excellent longevity and should continue to develop in the bottle for an additional eight to ten years.

1997 Grandpére Zinfandel

The fruit for our Grandpére Zinfandel is from one of the oldest known producing Zinfandel vineyards in America. This vineyard was planted somewhere in the mid to late 1850's and, amazingly enough has escaped the perils of prohibition and phylloxera - both of which doomed many early planted vineyards in California. These old vines provide us with low yields of tightly clustered (shot-berry size) Zinfandel grapes of deep, concentrated flavors. This vineyard is head trained with a larger canopy and its additional leaf coverage allows for the fruit to achieve the perfect balance of sugar ad true flavor ripeness. The fruit for this vintage was picked at 28° Brix and harvested on September 24, 1997.

After the two-week extended maceration period, which included 3 pumpovers per day to obtain greater extraction, the wine aged for 23 months in a combination of one to two year-old French and American oak barrels. Just prior to bottling in July of 1999, the wine went through minimal filtration for added clarity.

Related information:
 Alcohol: 15.3%   **pH**: 3.67
TA: 0.70 g/100 ml   **Bottling date**: July 1999
Production: 1457 cases   **Retail**: $30.00

Winemaker's Notes: Winemakers who specialize in Zinfandel production will attest to the belief that vine age has an important influence on wine quality. So, what are these magical qualities that the vine convey to the wine? Not simply extra ripeness or flavor intensity or tannin. In fact what makes Grandpére so special is an overall balance and harmony among the rich concentrated aromas, the silky texture and layers of flavor that are unmatched in other wines. You will find that the 1997 vintage of Renwood Grandpére is a classic example of what this renowned vineyard can provide - rich, full-bodied with excellent length and depth of flavors including spicy, black pepper, cedar, blackberries, cassis and anise.

1998 Fiddletown Zinfandel

The Fiddletown appellation is the oldest AVA in Amador County. The vineyards have a higher elevation, up to 1800 feet, and enjoy the perfect combination of hot days and cooler nights - more so than most Amador County vineyards. These growing conditions allow for maximum ripeness and balance in the fruit.

The 1998 was a cooler year which resulted in a later than usual harvest. The cooler weather, combined with some late rain in June contributed to smaller yields throughout Amador. The fruit from our three Fiddletown vineyards ripened evenly through the late summer months and was allowed to hang for maximum flavor-ripeness through the end of October.

After 30% whole berry fermentation - at cooler temperatures to extend skin contact for greater extraction - the wine was aged in 65% 2-4 year-old French and American oak and 35% new, deep toasted American oak for fifteen months.

Related information:
Alcohol: 14.0%   **pH**: 3.69
TA: 0.71 g/100 ml   **Bottling date**: April 2000
Production: 2620 cases   **Retail**: $25.00

Winemaker's Notes: Our 1998 Fiddletown Zinfandel has blueberry and cinnamon spice aromas that exemplify the unique characteristics of the Fiddletown AVA. This is a spicy Zin with nuances of bright wild berries and a delicate, lingering finish. It is deeply colored with a firm structure that allows the fruit to show through with full richness. These attributes make this wine perfect for early consumption or for cellaring an additional 4-6 years.

1998 "Old Vine" Zinfandel

Our "Old Vine" Zinfandel has become the flagship for all Renwood wines. It is a distinctive Zinfandel from eight of Amador County's oldest and most venerable vineyards. With an average age of at least 50 years, these head-trained, low-yielding sites provide us with a wonderful selection of even ripened, concentrated Zinfandel fruit that gives us great complexity and depth.

1998 provided us with smaller than usual yields due to the light rain that we received in late spring. However, the long mild summer that followed allowed for even ripening and resulted in smaller crops with full, intense flavor. The fruit was harvest in mid to late September and went through temperature controlled fermentation, with gentle pump-over regimens, to retain the full flavors and bright fruit characters. Following fermentation, the wine was aged in 30% new American oak barrels and 70% 2-3 year old barrels for 16 months.

Related information:
Alcohol: 14.0%   **pH**: 3.55   **TA**: 0.71 g/100ml
Bottling date: May 2000   **Production**: 8996 cases   **Retail**: $18.50

Winemaker's Notes: Our 1998 "Old Vine" Zinfandel has aromas of dusty berry, chocolate, white pepper underlined by subtle oak characters. The bright, brilliant berry and black cherry characters are highlighted by a firm, yet supple, mouthfeel and finish with a deep layer of vanilla and maple.

1997 Grandmére Zinfandel

Our "Grandmére" Zinfandel represents a blend of two of the finest vineyards that reside throughout the hillsides of Amador County's gold country. These two vineyards provide us with fruit from 60 and 80 year-old vines. We chose the name Grandmére (or Grandmother) because fruit from these vineyards result in a wine that is consistently warm and friendly, gracefully mature and immediately inviting.

A mild spring and long warm summer contributed to a textbook growing season for Amador County Zinfandel in 1997. The hand-selected fruit reached full maturity and was harvested in mid to late September. Following fermentation in open top stainless steel fermenters, the wine was aged for 18 months in a combination of French and American oak.

Related information:

Alcohol: 15.3%   **pH**: 3.59   **TA**: 0.70 g/100ml
Bottling date: July 2000   **Production**: 1647 cases   **Retail**: $25.00

W i n e m a k e r ' s   N o t e s : The 1997 Renwood Grandmére Zinfandel is big, bold, and ripe with bright red berry fruit and hints of lavender and anise on the nose, followed by explosive, dark spicy fruit and a well-rounded mouthfeel. It has a firm structure with lasting vanilla tones on the finish.

1997 D'Agostini Bros. Zinfandel

This vineyard designated Zinfandel is produced from one of the oldest growing Zinfandel vineyards in Amador County - the D'Agostini vineyard. This vineyard was planted in 1914 and has been tended to by the same family for 85 years. These old vines provide us with intense fruit that exclude classic "old world" characters in the wine.

The 1997 harvest conditions were excellent in Amador County and the D'Agostini Bros. vineyard, in particular, gave us fruit with increased intensity and depth from the crop which included 25% shot berries. The grapes were harvested on October 1,1997 and were fermented with a native yeast strain in large oak tanks. Following the maceration period of two to three weeks, the wine was aged in American Oak for 18 months in 50% two year-old and 50% three to four year-old barrels. The use of older oak allowed the fruit intensity not to be over-dominated by the oak nuance.

Related information:
Alcohol: 15.6%   **pH**: 3.52   **TA**: 0.62 g/100ml
Bottling date: July 1999   **Production**: 550 cases   **Retail**: $30.00

W i n e m a k e r ' s   N o t e s : The 1997 Renwood D'Agostini Bros. Zinfandel is a rich, complex Zin of black fruits, berries and cedar. Although this wine is drinkable now, added complexity will be obtained with additional five to seven years cellaring.

Renwood Winery Inc. also produces Barbera, Sangiovese, Syrah, Viognier, White Zinfandel, Muscato and an Ice Zinfandel.

R.H. Phillips

R.H. Phillips
26836 Country Road 12A
Esparato, CA 95637
Phone: 530-662-3215 Fax: 530-662-2880
Email: rhp@rphillips.com
Website: www.rhphillips.com

Since they first purchased their ranch in the Dunnigan Hill of California in 1946, the Giguiere Family and its R.H. Phillips Wine Company have traveled as undulating a roller-coaster as the land upon which they planted their vineyards. Against perceived wisdom and powerful odds, they grow grapes and produce premium wines on land historically devoted to sheep ranching and wheat farming. From the Ground Up is a phrase that truly describes this family and its wines.

R.H. Phillips, named in honor of their Grandfather - producer of fruit-focused, boldly-flavored classic and Rhône varietal wines - was founded by John, Karl and Lane Giguiere when they planted their first vineyard in 1981. It is the only winery, as well as the largest grape grower, in California's relatively new Dunnigan Hills Viticultural Area. The Giguieres have won praise for combing the philosophy of the small wineries just west of them in the Napa Valley with the economic advantages of the large wineries to their south in the Central Valley.

Tasting room open daily from 11-5

President: John Giguiere

Vineyard Operations: Karl Giguiere III

Winemaker: Barry Bergman

1997 Kempton Clark, Lopez Ranch

Appellation: Cucamonga Valley

Vineyard: Lopez Ranch

A mild winter with heavy spring rainfall was followed by a hot summer. Hot dry weather at the end of summer brought on the earliest harvest, August 6th, at the Lopez Ranch in recent memory, with high grape sugars due to the heat. Grapes were harvested from head-trained vines planted in 1915 from four vineyards that are dry farmed and yield only one-and-a-quarter tons per acre.

Composition: 100% Zinfandel

Vinification and Maturation: The juice was cold-soaked at 55° Fahrenheit for 24 hours prior to inoculation with D254 and fermentation in stainless steel tanks for six days at 85° F until dry and then pressed off the skins. The wines underwent malolactic fermentation with Viniflora Oenos direct inoculation and were then aged eight months, 70% in Tonnellerie Française Pennsylvania, Demptos and Sequin Moreau American oak, and 30% in Mercier Allier, Dargaud & Jaegle French oak, a combination of third and fourth year 225 liter barrels. The four lots were then blended and bottled.

Related information:
Alcohol: 14.5%
Residual sugar: Dry
pH: 3.46
Total Acidity: 7.3g/l
Bottling Date: September 8,9,and 10, 1998
Release Date: September 1998
Retail: $18.00

W i n e m a k e r ' s   N o t e s :
The Kempton Clark line, introduced in 1998, is devoted exclusively to heritage Zinfandels. The Giguieres named it in honor of their friend and neighbor, Kempton Clark, who in 1962 planted a 16-acre Zinfandel vineyard two miles from their ranch, to produce a wine for his Basque shepherds. His Basque Red, as it was dubbed, soon became a local legend. Today he is retired and since 1995 the Giguieres have been farming his vineyard.
A robust, spicy wine with plenty of ripe berry fruit for the Zin fanatic. Full-bodied with blackberry aromas, expansive grape flavors and a hint of black pepper spice, it has full tannins and good structure with a long, lingering finish.

1 9 9 7   K e m p t o n   C l a r k ,   M a d   Z i n

Appellation: California
Vineyard: 77% Cucamonga Valley Zinfandel 8% Dunnigan Hills Zinfandel 12% Paso Robles Petite Sirah and 3% Lodi Alicante Bouschet
A temperate growing season with beautiful weather throughout California that promoted very healthy berry set and canopies. Fruit developed slowly and matured with complex concentrated flavors and

good acidity. The Cucamonga grapes were harvested from head-trained, dry farmed vines that yield only 1¼ ton per acre. The Dunnigan Hills Zinfandel was produced from a variety of clones/field selections, on the Gable and Jones Ranches. The grapes were night harvested, at just over 24° Brix..

Vinification and Maturation: Harvest of the Zinfandels began in Cucamonga on August 6 and in the Dunnigan Hills on September 18. The juice was cold-soaked at 55°F for 24 hours prior to inoculation with D254. Fermentation, for six days at 85°F, was in stainless steel, where the wines remained in skin contact for another 13 days. They underwent malolactic with direct inoculation and were then aged eight months in oak.

Related information:
Alcohol: 14.0%
Residual sugar: Dry
pH: 3.37
Total Acidity: 7.7g/l
Bottling Date: December 9, 17, and 18, 1998
Release Date: January 1999
Retail: $12.00

Winemaker's Notes:
MAD (acronym for Multiple Appellation Designate) is part of Giguieres' Kempton Clark line, introduced in 1998, which is devoted exclusively to heritage Zinfandels. A spicy, jammy, fruit-forwarded Zinfandel highlighted with toasty American oak. Paso Robles Petite Sirah adds bright fruit flavors and robust tannins while the Alicante Bouschet brings rich, deep color.

R.H. Phillips also produces Chardonnay, Cabernet Sauvignon, Syrah, Merlot, Malbec, Viognier and Sauvignon Blanc.

Ridge Vineyards

Ridge Vineyards
17100 Monte Bello Road
P.O. Box 1810
Cupertino, CA 95015
Phone: 408-867-3233 Fax: 408-868-1350

In 1959, the Ridge Partners re-opened the old Monte Bello winery located at 2600' in the Santa Cruz Mountain and built in the 1880's. The first of the new estate wines came from Cabernet vines re-planted in the 1940's. In 1964, Ridge made its first Zinfandel from century old vines lower on the ridge and in 1966 its first Geyserville Zinfandel from the 19th Century vines on the Trentadue ranch in Sonoma County. It became clear that although the limestone soils and very cool climate of the Ridge were ideal for the Bordeaux varietals and a small amount of Chardonnay, it was too cool to fully ripen Zinfandel year in and year out. After working with at least 40 different single Zinfandel vineyards, Ridge leased Geyserville and purchased Lytton Springs, two of the finest, most consistent properties.

Tasting Room is open Saturday and Sunday from 11-3 at the Ridge location of 17100 Monte Bello Road, Cupertino. The Sonoma Station facility has various hours depending on the season. For the Sonoma Station, please call ahead at 707-433-7721.

Owner: Akihiko Otsuka   **Winemaker and CEO**: Paul Draper

Winemaking Philosophy: What distinguishes wine from all other alcoholic beverages and gives it a meaning beyond its mind-altering properties is its connection to nature. In the finest wines it is the distinctive quality and character of the fruit from a particular piece of ground (terroir) that determines the quality and individual character of the wine. The grape itself has in or on it all that is necessary for it to become wine: sugar, acid, tannin in the seeds, color in the skins and yeast on any tiny break in its surface. Because of this, from the beginning of Western civilization, it has been the central symbol for transformation both physical and spiritual. It is part of both Christian and Jewish ritual for this reason. In order that our wines carry this meaning, we have selected a hand full of vineyards from the more than forty with which we have worked. The distinctive character from each is consistently so clear and of such quality that the wine does not require blending with other vineyards nor any physical or chemical manipulation to be of the highest quality. Our vineyards and our gentle handling of the fruit produce full bodied wines that can handle their firm structure and still be sensuous. They are aged exclusively in air-dried American oak.

1998 Ridge California Geyserville

Appellation: Sonoma County

Vineyard Designation: Geyserville Vineyard

The Geyserville Vineyard, owned by Leo and Evelyn Trentadue, is located in Sonoma County, 3 miles south of the town of Geyserville on the western edge of Alexander Valley. The vineyard consists of 15 acres of 107 year-old Zinfandel, 12 acres of 36 year-old Zinfandel, 5 acres of 117 year-old mixed blacks, 5 acres of 17 year-old Petite Sirah and 7 acres of 107 year-old Carignane.

The soils are gravely loam yielding 1 to 3 tons per acre. At an elevation of 200 feet, the vineyard experiences occasional morning fog and warm days with frequent evening breezes.

Composition: We first chose to work with the oldest parcels of this vineyard in 1966. We were attracted by the gravely soils and the flavor and intensity of the grapes from these low yielding vines. The parcels we lease or contract for are well above the rich, river bottom soils to the east. The various parcels of the vineyard include plantings of Petite Sirah, Carignan, and a small amount of Mataro (Mourvèdre) and a mixed planting of all of the above. The age of the vines vary from ten years to just short of 120 years. All vines are head trained without any trellising.

Vinification: The majority of the Zinfandel parcels are fermented using a submerged cap with the grapes held just below the surface of the liquid.They are pumped over daily using a gentle irrigation device. Wines are pressed to taste, almost always at dryness, typically after eight to twelve days on the skins. All the primary and secondary fermentations are carried out on the yeast and malolactic bacteria that are on the grapes as they arrive from the vineyards. Only with Petite Sirah are the press wines kept separate. Each parcel within the vineyard is fermented separately and after pressing, the task of assembling the final vineyard wine begins. The wines are blind tasted and the most intense, but also most typical of the vineyard are selected for the final wine. This process is often not completed until a year later - four to six months before bottling.

Maturation: After a natural malolactic, complete typically in late November, the wine is racked to air-dried, American oak barrels.

Usually about twenty percent goes into new barrels, the rest into one-two-year-old and older oak. The wine is racked off the sediment after one to two months and this is repeated every three to four or five months until bottling. No manipulation such as spinning cone or centrifuging is used and no filtration at any level is used until bottling. The wines are never membrane sterile filtered unless there is an active secondary yeast fermentation and this has not been necessary for over ten years. The wines are normally released within three to six months of bottling and are enjoyable at that time although they improve considerably within twelve months of release. The fruit is most intense within seven years of vintage at which time the dominant fruit typically begins to diminish and secondary flavors from the oak and bottle aging come to the fore. At approximately fourteen years, a well structured Zinfandel in a fine vintage develops complex, almost exotic characteristics as the ripe fruit flavors and firm structure evolve. Most Zinfandels are at their most sensuous in those first five to eight years.

Related information:
Alcohol: 14.1%   **Residual sugar**: 0.0451%
Brix at Harvest: 24.8   **Harvest Date**: 10/2-11/13/98
Bottling Date: January 2000
Release Date: Fall 2000
Production: 11,169 cases sold at winery and distributed.
Retail: $27.50

1998 Ridge California Lytton Springs

Appellation: Sonoma County
Vineyard Designation: Lytton Springs Vineyard.
Located on the eastern edge of the Dry Creek appellation, Lytton Springs, East and West, is owned by Ridge Vineyards. Lytton East consists of 42 acres planted in 1892 to Zinfandel, Petite Sirah, Grenache and Carignane. Lytton West is a younger vineyard, with 33 acres of 45 year-old Zinfandel, Grenache, Carignane and another 27 acres of 3 to 10 year-old Zinfandel, Petite Sirah, Grenache and Mataro. At an elevation of 80 to 160 feet the soils vary within each vineyard with a predominance of gravely clay and gravely clay loam on the hillsides. Exposure is southeasterly with fog in the morning, warm sunny afternoons and breezes late in the evening.
Composition: This series of parcels on our Lytton East and Lytton West properties on Lytton Springs Road lies on rolling hills and bench land

between Alexander Valley and Dry Creek Valley. It is within the Dry Creek appellation. We first harvested the grapes from the now 108 year-old vines on the eastern parcels in 1972. The clay and gravel soils produce a wine that combines intense fruit with an earthy, robust structure. The vines vary in age from 3 to 108 years and include Petite Sirah, Grenache, Mataro and Carignane.

Vinification and **Maturation** For all Ridge Zinfandels vinification and maturation are virtually the same.

Related information:
Alcohol: 14.3%   **Residual sugar**: 0.0857%
Brix at Harvest: 25   **Harvest Date**: 9/16-10/22/98
Bottling Date: December 1999   **Release Date**: Spring 2000
Production: 8,226 cases sold at winery and distributed.
Retail: $27.50

1998 Ridge California Zinfandel York Creek

Appellation: Napa Valley
Vineyard Designation: York Creek Vineyard
Located on Spring Mountain on the western border of Napa County, York Creek Vineyard is owned by Fritz Matag. At an elevation of 1600 to 1950 feet above sea level, this 12 acre vineyard consists of 22 to 31 year-old Zinfandel producing 2.5 tons per acre. The soils are well-drained boomer gravely loam and Felton loam with the climate cool and mountainous.

Composition: High on Spring Mountain, above St. Helena, the remnant parcels of old Zinfandel and Petite Sirah were expanded in the late sixties. We made our first York Creek Zinfandel in 1975. Black pepper spice, berry fruit and firm structure set these wines apart from the northern Sonoma Zinfandels. They are more Cabernet-like in structure.

Vinification and **Maturation** For all Ridge Zinfandels vinification and maturation are virtually the same.

Related information:
Alcohol: 14.9%   **Residual sugar**: 0.1211%
Brix at Harvest: 26.1   **Harvest Date**: 10/27-10/29/98
Bottling Date: April 2000   **Release Date**: Fall 2000
Production: 3,562 cases sold at winery and distributed.
Retail: $27.50

1998 Ridge California Zinfandel Pagani Ranch

Appellation: Sonoma County
Vineyard Designation: Pagani Ranch.
Owned by the Pagani Family the 38 acre vineyard is primarily Zinfandel, part of which is interplanted with Petite Sirah, and includes six acres of Mataro and two acres of Alicante Bouche as separate parcels, and produces 1.5 to 2.5 tons per acre. Clay loam with old creek-beds meander throughout the rolling, east facing slopes. Located in the Northern Sonoma Valley the vineyard experiences cool, foggy mornings and warm days.
Composition: The oldest blocks of this vineyard were planted around the turn of the century and the youngest just prior to prohibition fifteen or twenty years later. It is our one vineyard in which all of the vines are very old as there has been no replanting even to replace those parcels destroyed by deer. The intensity of these old, low yield vines provides a distinctive spice and complex fruit when fully ripe. Besides the Zinfandel, there is Mataro (Mourvèdre), Petite Sirah and Alicante Bouschet as well as very small percentages of other lesser known varieties.
Vinification and **Maturation** For all Ridge Zinfandels vinification and maturation are virtually the same.

Related information:
Alcohol: 14.2%   **Residual sugar**: 0.0644%
Brix at Harvest: 25.3   **Harvest Date**: 10/18-10/21/98
Bottling Date: March 2000   **Release Date**: Fall 2000
Production: 2,976 cases sold at winery and distributed.
Retail: $27.50

1998 Ridge California Paso Robles

Appellation: Paso Robles
Vineyard Designation: Dusi Ranch.
Owned by Benito Dusi, this 40 acre vineyard is located three miles south of Paso Robles. Rocky and gravely, with some areas of light soil, 81 year-old Zinfandel produces 1.5 tons per acre. At an elevation of 760'

the vineyard experiences hot days and cool nights with full exposure in all directions.

Composition: We first made this wine in 1967 and have made it every year from 1976 onwards. The full berry fruit and regional spice are carried by a structure that softens within two or three years of age to make it one of the most approachable of our Zinfandels. All but a few of the vines are now seventy nine years old and include a tiny amount of young Petite Sirah.

Vinification and **Maturation** For all Ridge Zinfandels vinification and maturation are virtually the same.

Related information:

Alcohol: 14.9%	**Residual sugar**: 0.0250%
Brix at Harvest: 26.	**Harvest Date**: 10/01-10/15/98
Bottling Date: December 1999	**Release Date**: Spring 2000

Production: 2858 cases sold at winery and distributed.
Retail: $22.50

1998 Ridge California Zinfandel Sonoma Station

Appellation: Sonoma County
Vineyard Designation: No designation.
A blend of Dry Creek, Russian River, Alexander and Sonoma Valley vineyards, Ponzo, Mazzoni, Lyttton West and Pagani are the source of the Zinfandel, Petite Sirah, Carignane and Alicante Bouschet that create the Sonoma Station Zinfandel.

Composition: This wine is the successor to the colorful grape cluster label of the former Lytton Springs facility on Chiquita Road outside of Healdsburg. The same wine from the Mid-Way Ranch is the basis for the blend that typically includes Zinfandel, Petite Sirah, Carignan and Alicante Bouschet. It comes from the same group of vineyards included over the past six years. This is one of two wines that we ferment primarily at Sonoma Station instead of the Monte Bello estate winery.

Vinification and **Maturation** For all Ridge Zinfandels vinification and maturation are virtually the same.

Related information:

Alcohol: 13.7%	**Residual sugar**: 0.0359%
Brix at Harvest: 24.2	**Harvest Date**: 10/02-10/28/98
Bottling Date: December 1999	**Release Date**: Spring 2000

Production: 8454 cases sold at winery and distributed.
Retail: $18.00

Ridge Vineyards also produces the Monte Bello Cabernet Sauvignon, Cabernet Sauvignon Santa Cruz Mountains, Merlot Santa Cruz Mountains, Chardonnay Santa Cruz Mountains, and Petite Sirah York Creek.

Robert Mondavi Winery

Robert Mondavi Winery
7801 St. Helena Highway
Oakville, CA 94562
Phone: 1-888-RMONDAV
Website: www.robertmondaviwinery.com
Email: info@robertmondaviwinery.com

The winery was founded in 1966 in the heart of Napa Valley and is the flagship winery of the Robert Mondavi company. In early 1999, the winery undertook the To-Kalon Project, the first major refurbishing in its 33 year history, scheduled to be completed in 2001.

Owner: Robert G. Mondavi family and shareholders. (traded on NASDAQ:MOND)

Winemakers: Tim Mondavi, managing director and winegrower and Genevieve Janssens, Director of Winemaking.

Viticulturists: Mitchell Klug

Vice President and GM: Clay Gregory

Tasting room open daily from 9 a.m. to 5 p.m. daily except Easter, Thanksgiving, Christmas and New Year's day. Tours available. Please contact the winery for details on a variety of tastings and tours available for visitors.

Winemaking Philosophy:

We continue to seek out old-clone Zinfandel grapes for our Robert Mondavi Winery Napa Valley Zinfandel. Our goal is to produce a "Meridional-style" wine…in the style of Southern France or Italy…emphasizing freshness and vibrancy with intense fruit characters of ripe berry, plum and spice, firm structure and moderate tannin. We want the wine to have complexity from barrel aging, but without pronounced new oak character.

1998 Zinfandel

Vintage:

In Napa Valley, as throughout California, 1998 was an unusual year for weather. El Niño gave us a growing season punctuated with extremes -

twice the normal rainfall, 18 days of frost, cold weather at bloom and record heat spells. Yields for some varieties were down because of the wet weather during bloom and berry set. A period of excessive summer heat caused sunburn in some vineyards, necessitating thinning to maintain high quality. But then, as harvest approached, the weather became increasingly better, and we had near perfect, mild, sunny days for ripening. The small loose clusters had sufficient hang time on the vines to create vibrant flavors and natural balance in the wines. Overall, the 1998 harvest, one of the latest of the decade, turned out to be of excellent quality.

V i n e y a r d s :
The deep flavors of our Zinfandel are from old clones of Zinfandel vines that produce low yields of small, intensely flavored grapes. We have sought out old clone, head trained, dry farmed vineyards from the entire length of Napa Valley to give depth and complexity to our 1998 Zinfandel. This wine combines 40-80 year old vines with new plantings that were grafted with budwood from old, exceptionally high quality vineyards.

C o m p o s i t i o n :
91% Zinfandel, 5% Charbono, 4% Petite Sirah

V i n i f i c a t i o n   a n d   M a t u r a t i o n :
At the winery, the grapes were gently destemmed and crushed. We ferment the must with a cultured yeast strain selected to enhance the forward fruit character of Zinfandel. Depending on the specific vineyard's personality, we gave the wine an average of 16 days extended skin contact to soften tannins and heighten varietal character. After the wine was pressed and transferred to small oak barrels, it underwent a slow malolactic fermentation to enhance the round, complex mouthfeel. We aged the blend in thin-staved, chateau-style French oak barrels for eight months to mature and integrate flavors.

Related information:
Alcohol: 14.6%
Brix at Harvest:23.8
Residual sugar: Dry
pH: 3.76   **TA**: 0.56%
Harvest Date: Harvested over a period of days in late September 1998
Bottling Date: August 1999   **Release Date**: Scheduled for June 2000
Retail: $19.00

W i n e m a k e r ' s   N o t e s :
Our 1998 Napa Valley Zinfandel displays ripe tiers of briary berry, plum and black cherry, with pepper and oak nuances weaving throughout the wine. The

vibrant fruit flavors and velvety texture of this Zinfandel are attributed to gentle handling throughout the winemaking and aging in French oak barrels.

Food pairing: Our 1998 Zinfandel is delicious with bold flavors of grilled meats, braised short ribs and hearty stews or vegetable dishes. The effusive, peppery spice character is also a great companion to pasta dishes and pizzas with sun dried tomatoes, eggplant, fresh garlic and basil.

Robert Mondavi Winery also produces Cabernet Sauvignon, Chardonnay, Fume Blanc, Merlot, Pinot Noir, and Moscato d"Oro.

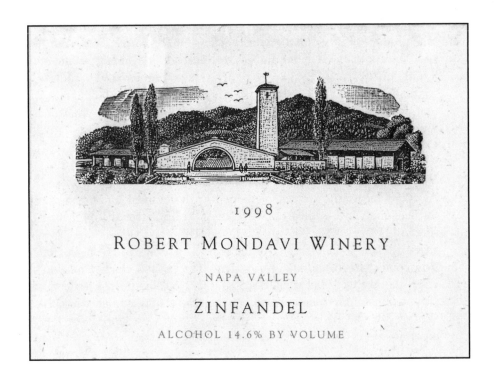

Rochioli Vineyard and Winery

Rochioli Vineyard and Winery
6192 Westside Road
Healdsburg, CA 95448
Phone: 707-433-2305   Fax: 707-433-2358

Premium grape growers since the 1930's, Rochioli family's commitment to quality results in award-winning wines.

Tasting Room open daily from 10:00am to 5:00pm, December - January 11:00 a.m. to 4:00 p.m. daily. Picnic area overlooks the vineyard. Continuous art shows in tasting room.

Owner: Joe and Tom Rochioli   **Winemaker**: Tom Rochioli

Winemaking Philosophy: We try to bring out the best the vineyard has to offer. This usually requires very little tampering with the vines, if your vineyard and grapes are sound. The trick is to figure out what method to use to bring out the best. There is no real recipe to follow, but the goal is to create wine that all will enjoy.

1998 Sodini Vineyard Zinfandel

Vineyard: The Sodini Vineyard is located on the Eastern side of the Russian River Valley, just south of Healdsburg on Limerick Lane. These vines were planted in 1905.
Composition: 100% Zinfandel
Vinification and Maturation: Traditional winemaking, small tank, hand-punched. Aged in 60 gallon French oak for 16 months.

Related information:
Alcohol: 14.5%   **Residual sugar**: Dry
Total acidity: 0.65g/100ml   **Harvest Date**: 10/12/98
Bottling Date: February 2000   **Release Date**: 3/01/00
Production: 250 cases sold at winery
Retail:$25.00

Winemaker's Notes: Spicy berry fruit combines with the blackberry character of old vine Zinfandel. This wine is full bodied with rich flavors of raspberry and vanilla that finishes full and balanced.

1998 Rochioli Estate Vineyard

Vineyard: This small vineyard is located on our hillside property facing to the southwest where it receives the full sunlight necessary for complete ripening during the growing season. The vines were planted in 1969 and yield only 1.5 tons per acre.

Composition: 100% Zinfandel

Vinification and Maturation: Traditional winemaking, small tank, hand-punched. Aged in 60 gallon French oak for 16 months.

Related information:

Alcohol: 14.5%	**Residual sugar**: Dry
Total acidity: 0.65g/100ml	**Harvest Date**: 10/17/98
Bottling Date: February 2000	**Release Date**: 3/01/00

Production: 100 cases sold at winery

Retail:$28.00

Winemaker's Notes: Ripe aromas of blackberry jam, vanilla bean and a touch of pepper combine with rich berry flavors.

Rochioli Vineyard and Winery also produces Estate Sauvignon Blanc, Estate Chardonnay, Estate Pinot Noir.

Rocking Horse

Rocking Horse
P.O. Box 5868
Napa, CA 94581
Phone: 707-226-5555   Fax: 707-255-1506

Rocking Horse is a small quality producer of Napa Valley Ultra Premium red wines. When established in 1989, production was limited to 600 cases. Born from the desire to produce the absolute best wines from legendary growing regions the Doran and Zealear families began their quest moving quietly from a backyard bottling line to their current 6,000 case production. Extreme care is taken to marry the proper vineyard sight and appellation with the appropriate varietal. Trademark wines are typically "made in the vineyard" and picked at optimum ripeness to insure a silky full mouth feel and dark crimson purple hue. Each wine is true to its varietal character and offers many layers of complexity. Properly aged wines will be rewarded by cellaring.

The name Rocking Horse was developed over time from an in-house collection of art and sense of belonging. Memorable in nature yet understated by design, the name touches the heart of Americana and lingers on…well after the evening's wonderful meal.

Currently there is no tasting room or tours.

Owners: Jeff and Nancy Doran   Brian and Candida Zealear

Winemakers: Jim Moore "the artist" and Steve Lagier "the scientist"

Winemaking Philosophy:

The philosophy of our winemaking is very complex in nature and requires that attention be paid to every detail. There are literally thousands of decision made throughout the winemaking experience and includes both the science and the art of winemaking. However, extracting varietal character is the primary focus of our winery from canopy management and pruning to sorting and barrel types. A final blend typically evens out any broad variations offered by our other partner, Mother Nature, which can include the wrath of rain, wind and snow; mountain lions, bears, rabbits, birds, and deer; phylloxera, sharp shooters and other insects.

The Vintage:

1998 was a very difficult harvest for Zinfandel. Many vineyards were affected by rot and mildew. Those of us that set high standards for harvest, did a great deal of hand sorting and pruning. Sugars were allowed to escalate despite the difficulty of harvest to obtain a balance between acid and pH. Our wines are typically blended to achieve their best. We are extremely pleased with the results, despite the fact that 1998 was such a difficult harvest.

1998 Napa Valley Zinfandel

This vintage was a 4 vineyard blend consisting of 91% Zinfandel and 9% other red blending varietals used to maximize flavors and balance the wine. Two vineyards were from Howell Mountain giving the wine a bold yet characteristic pepper and spice orientation while the balance of fruit was from the broader Napa Valley. The wine shows typical notes of pepper, clove, hints of cherry and spice, brilliantly clear, crimson in color, exhibiting an elegant, full mouth feel and lingering finish.

The wine was cellared for 16 months with 15% new American oak barrels and blended to achieve a consistent harmonious finish.

Related information:
Alcohol: 15.2%
Residual sugar: Dry
pH and TA: 3.70 / 0.62
Harvest Date: 10/11/98
Bottling Date: Feb 2000
Release Date: 5/1/00
Production: 2550 cases produced
Retail: $18.00

Rosenblum Cellars

Rosenblum Cellars
2900 Main Street
Alameda, CA 94501-7522
Phone: 510-865-7007 Fax: 510-865-9225   Email: Rosenblumcellars.com

Rosenblum (pronounced "rows en bloom") Cellars is located in the historic Todd Shipyard on the island of Alameda. The cooling breezes from the San Francisco Bay give us natural air conditioning and the perfect climate to produce and cellar fine wines. Established and bonded in 1978, Rosenblum Cellars has consistently produced award winning wines. Our grapes come from over 30 different vineyards throughout California.

Tasting Room open daily from 12:00 to 5:00 pm. New-release Wine Club. Please contact the winery about all the special events they hold throughout the year and the facility rental.

Proprietor and Winemaker: Kent Rosenblum

Winemaking Philosophy: Rosenblum Cellars makes mostly red wine. Seventy-five percent of my production goes toward making 14 or 15 different Zinfandels per year. Each wine I produce is made with a minimal amount of machinery and manipulation. The grapes are pampered throughout the winemaking process. Great care is taken to produce rich red wine with lots of varietal fruit characteristics, great complexity and a soft mouthfeel! My goal at Rosenblum Cellars is to continue to search out premium vineyards, produce luscious, user friendly wine at reasonable prices utilizing natural, historical methods and the finest oak barrels.

Philosophy and Method:
Starting with sound, ripe grapes which are from a mountain and older, head-pruned vineyards. The grapes from these special vineyards are hand harvested and quickly brought to the winery for crushing.
Using a labor intensive, small batch, hand punch down system for fermentation, this is a traditional European technique and is now used by only a handful of small California wineries. It allows us to produce a wine rich in bouquet, flavors and color, but with soft tannins and lower alcohol.

Utilizing high quality, small oak cooperage to age our wines. These barrels come from some of the finest forests in both France and America. Minimal handling while the wines are in the cellar, and often bottled without fining or filtration.

Our promise to you is top quality, natural wine at a fair price. Stop and Smell the Rosenblums.

1998 Sonoma County Zinfandel St. Peters Church

Vineyard: St. Peters Church, Cloverdale, Sonoma County
This vineyard, planted in 1898, is located within the Cloverdale city limits behind St. Peter's Catholic Church. The soil is black loam with gravel, planted with about 60% Carignane and 40% Zinfandel vines. Typical yield is 1 to 2 ½ tons per acre.
Composition: 78% Zinfandel and 22% Carignane.
Vinification and Maturation: The grapes for this wonderful old vineyard were hand picked September 15, 1998. After being transported to the winery, they were crushed and fermented in small open top bins utilizing a twice daily hand punch down technique. The fermentation was quick and warm, followed by a 3 day post fermentation maceration and gentle pressing. The Carignane was fermented separately and the two wines were blended in December 1999. The wine was aged in 100% new French coopered American oak for twenty months, and racked four times before bottling in May 2000. This is one of four small production wines bottled in a special tall gold painted bottle.

Related information:
Brix at Harvest: 25.2   **Harvest Date**: September 1998
pH: 3.38   **Total Acid**: .78gm/100ml
Bottling Date: 5/2000   **Release Date**: July 2000
Production: 200 cases available at the winery only

Winemaker's Notes: We feel that this wine reflects divinity of the vineyard. The combination of Zinfandel and Carignane tastes like it was made in heaven. Ripe black currant and boysenberry fruit pervade the bouquet followed by flavors of ripe dark cherries and currants with elements of chocolate, licorice and menthol.

1998 Samsel Zinfandel
Maggie's Reserve

Vineyard: Samsel Vineyard, Sonoma County.
The Samsel vineyard sits on a sloping plain just west of Arnold Drive in mid-Sonoma Valley, two miles south of Glen Ellen. It is planted with 95% Zinfandel and 5% Alicante Bouschet. The vines planted at the turn of the century, are head-pruned, dry farmed and yield about 2 to 3 tons per acre. Hot days and moderately cool nights during the growing season mark the climate in this region.

Composition: 87% Zinfandel, 8% Petite Sirah and 5% Alicante Bouschet

Vinification and Maturation: Ripe grapes were hand harvested in October 1998. Fermentation occurred in ¾ ton bins, with the cap punched down by hand two times daily for 9 days. 10% whole clusters were retained during the fermentation process. After a 3 day post-fermentation maceration the must was pressed to 60 gallon oak barrels (10% new American, 20% older American and 70% new French oak barrels). The wine was aged for 16 months and bottled unfined and unfiltered in March 2000.

Related information:

Brix at Harvest: 25.4	**Harvest Date**: October 15, 1998
pH: 3.61	**Total Acid**: .64gm/100ml
Bottling Date: 3/2000	**Release Date**: April 2000
Production: 625 cases	

W i n e m a k e r ' s   N o t e s : The aromas of wild raspberry, blackberry and a touch of eucalyptus and vanilla come all wrapped up with the subtle hints of herbs and cinnamon. The flavors are even more extracted and compelling than the aromas. The flavors are a cornucopia of rich raspberry, blackberry and bing cherries with hints of cloves and eucalyptus. The finish on the wine is so long and intense that it will make you wonder if life could get any better.

Vintners Cuvee Zinfandel

Appellation: California: The source of fruit changes with each vintage.
Composition: Zinfandel with a blend of other red varieties.
Vinification: Sterile filtered which tends to make the wine a bit smoother and readily drinkable upon release.

W i n e m a k e r ' s   N o t e s : The goal of our cuvee program is to produce a very friendly, drinkable, slightly light style Zinfandel at a very affordable price. The base of this wine is often not our best vineyard-designated Zinfandels, but Zins of good quality. To aid the consumer, we have simply designate each successive blend with a Cuvee number, so it can be specifically sought out without regard to a vintage date. A great Zinfandel produced to be drunk with pasta and pizza. It is a very soft yet full mouth filling Zinfandel with an emphasis on zesty fruit and moderate oak.

1998 San Francisco Bay Contra Costa County Reserve, Pato Vineyard

Vineyard Designation: Pato Vineyard, Oakley, Contra Costa County
This magnificent old vineyards sits at the foot of beautiful Mt. Diablo in the San Francisco Bay region town of Oakley. The old head pruned vines were planted between 1897 and 1902 in the typically sandy soil of the delta region by the Pato Family who have tended the vines ever since.
Composition: 98% Zinfandel 2% Mourvèdre
Vinification and Maturation: This wine is produced using cold soak tank fermentation with 15% whole clusters and 2 days of post fermentation maceration. After a gentle pressing the resulting wine was settled for 10 days and transferred into 60-gallon oak barrels. The wine was aged for 17 months in 20% French, 80% American, 50% new and 50% 1 year old. It was racked three times and bottle unfined with a gentle filtration in April 2000.

Related information:
Brix at Harvest: 25.4   **Harvest Date**: October 1998
pH: 3.62   **Total Acid**: .78gm/100ml
Bottling Date: 4/2000   **Release Date**: May 2000
Production: 500 cases

W i n e m a k e r ' s   N o t e s : This special lot of Pato Vineyard grapes was harvested at a higher Brix level to enhance the concentration and potential extraction. The resulting wine has a rich black cherry and spicy herb bouquet with hints of licorice and chocolate. The flavors are ripe black cherry with hints of currants, hazelnuts and bacon.

1998 Dry Creek Valley Zinfandel Rockpile Road Vineyard

Vineyard: Rockpile Road Vineyard, Dry Creek Valley
Jack Florence, Proprietor
This small hillside vineyards sits high above Lake Sonoma with a great view of the lake and upper Dry Creek Valley. The unique location and the fact this vineyard is planted to a rare clone of old vine Zinfandel makes this wine a very special creation. A small block of Petite Sirah is just coming into production, which is also a fine selection from a 100-year old clone. The name Rockpile Road is appropriate as the fence that borders Rockpile Road is make from rocks pulled from the vineyard. The soil is a volcanic loam with red clay and gravel.
Composition: 94% Zinfandel 6% Petite Sirah
Vinification and Maturation: The grapes were hand-harvested in early October 1998 and crushed into small fermenters. The cap was punched into the fermenting juice at least twice daily. A six-day post fermentation maceration followed. After a gentle pressing, the wine was aged in 60 gallon oak barrels, 40% French and 60% American for 16 months. The wine was racked four times and lightly fined and filtered before being bottled in February 2000.

Related information:
Brix at Harvest: 24.8   **Harvest Date**: October 1998
pH: 3.53   **Total Acid**: .79gm/100ml
Bottling Date:2/2000   **Release Date**: March 2000
Production: 1200 cases

Winemaker's Notes: The bouquet of this rich, extracted wine shows ripe black cherry and black currant with hints of vanilla, black pepper and menthol. The flavors are ripe black cherry, currant and creamy spice with hints of anise. This rich Dry Creek Zin will nicely accompany hearty pasta and grilled marinated salmon. It might even be considered for a clambake complete with piles of rocks and seaweed...make sure you bring along the rocky road ice cream.

1998 Napa Valley Zinfandel RustRidge Vineyard

Vineyard :This beautiful vineyard sits at a 1200 foot elevation in the rolling hills in the eastern Napa Valley. It is a small sub-appellation of Napa known as Lower Chiles Valley. The soil is gravely volcanic loam. The vines were planted in the 1970's, with five different Zinfandel clones making up the vineyard mix.

Composition: 100% Zinfandel

Vinification and Maceration: The ripe grapes were harvested when the flavors of the skins matched maturity. They were picked in three different lots and all fermented in small stainless steel tanks with a cold soak for 3 days, then a warm fermentation with temperatures up to 88°F ensued. It was pressed and settled before being transferred into 60 gallon oak barrels, 20% French, the remainder being American oak from four different coopers; Seguin Moreau, Nadalie, Radoux, and Demptos. The wine was aged for 16 months.

Related information:

Brix at Harvest: 25.2	**Harvest Date**: October 1998
pH: 3.41	**Total Acid**: .84gm/100ml
Bottling Date: 4/2000	**Release Date**: June 2000
Production: 1100 cases	

W i n e m a k e r ' s   N o t e s : This is a very lush and structured wine exhibiting a bouquet of ripe currants and spicy boysenberries with elements of smoked game and black pepper. The flavors are ripe black cherry and black berry with clean smoky elements and hints of chocolate. It should nicely accompany roasted goose stuffed with wild rice, hearty stews, rich pastas and grilled salmon.

1998 Annette's Reserve, Redwood Valley Rhodes Vineyard

Vineyard: Annette and Richard Rhodes' estate in Redwood Valley, Mendocino.

This vineyard encompasses about 60 acres of sloping red loam and shale in the eastern Redwood Valley of Mendocino County. 35 acres are planted to 42 year-old head pruned Zinfandel vines, with the balance planted to Carignane and Petite Sirah. Production is an average of 2-3 tons per acre.

Composition: 78% Zinfandel 15% Carignane 7% Petite Sirah

Vinification: The varietals were harvested separately at the peak of ripeness and produced individually using our traditional hand punch down technique and extended maceration. The best blend of the wine was determined in February 1999 and this blend was subsequently aged in 80% American oak (40% new) and 20% French oak for 19 months. The wine was racked four times before bottling in May 2000, unfined and unfiltered.

Related information:

Brix at Harvest: 24.2	**Harvest Date**: October 18-20, 1998
pH: 3.38	**Total Acid**: .67gm/100ml
Bottling Date: 5/2000	**Release Date**: July 2000
Production: 1800 cases	

W i n e m a k e r ' s   N o t e s : This beautiful wine comes from the Annette and Richard Rhodes' estate in Redwood Valley - a blend of 78% Zinfandel, 15% Carignane, and 7% Petite Sirah from the same ranch. This wine is a scholarly Zinfandel - a Rhodes scholar, with deep red color and an exotic bouquet of blackberries with evidence of rose petal, vanilla and mulled spice. The flavors are rich black cherry with hints of chocolate, black olives and cracked pepper.

1998 Paso Robles Zinfandel
Richard Sauret Vineyard

Vineyard: Richard Sauret's vineyard is located on a steep hillside in the Santa Lucia mountain, two miles west of Highway 101. Its elevation is the highest within the Paso Robles viticultural area. The soil consists of red clay and blue shale with some gravel and loam. The vines are 36 years old, head-pruned and yield on average three tons per acre. The climate during the growing season is generally hot, but occasional maritime influences bring cool air into the area.

Composition: 98% Zinfandel 2% Petite Sirah

Vinification and Maturation: Perfectly ripe grapes were harvested in early October, transported to Alameda and crushed into 4,000 gallon fermenters with 10% whole clusters and natural yeast. After holding the must at 50°F for three days, it was warmed to 72° for the start of fermentation. The caps in the fermenting bins were intensively sprayed twice daily and reached 91°F. A post fermentation maceration of three days ensued before a gentle pressing. The resulting wine was aged in 60 gallon oak barrels with 20% new American oak, 40% shaved French oak, and 40% one-year-old French oak for 16 months. It was gently filtered and bottled in February 2000.

Related information:

Brix at Harvest: 24.8   **Harvest Date**: October 2, 1998
pH: 3.53   **Total Acid**: .79gm/100ml
Bottling Date: 2/2000   **Release Date**: April 2000
Production: 4250 cases

W i n e m a k e r ' s   N o t e s : This is a rich and exotic fruit packed Zinfandel. A bouquet of ripe black currant and blackberry with smoky ripe cherry essences is followed by flavors of currant and creamy vanilla with hints of briary spice and cracked black pepper. This wine, in our opinion, is one of the best Paso Robles Zinfandels produced from Richard Sauret's vineyard. We expect it to age gracefully for 8 to ten years, but the opportunity to enjoy it in its youth shouldn't be missed. Pair it with Asian tea-smoked Duck, grilled Rack of Lamb, or a hearty polenta.

1998 Napa Valley Reserve Zinfandel George Hendry Vineyard

Vineyard: George Hendry Vineyard, Southern Napa Valley
The George Hendry Vineyard is located in the rolling hills on the southeastern foot of Mt. Veeder at 200 to 300 foot elevations. The Zinfandel was planted in 1976 on the uppermost eight acres of hillside in well-drained Boomer series soil. The UCD #2 clone Zinfandel is cordon trained on a 4-wire trellis system with 10 x 12 spacing. The vines are trained so that the bunches are evenly spaced and exposed to the sun. Yield is 4-4 ½ tons per acre with medium sized bunches sporting thick flavorful skins. Viticultural techniques include shoot positioning, cluster thinning, leaf pulling and cluster positioning to achieve a uniformly ripe bunch of fruit. Vineyard temperatures range into the high 90's.
Composition: 100% Zinfandel
Vinification and Maturation: The harvested grapes were crushed into ¾ ton fermenters with 15% whole cluster fruit. After soaking for three days, the wine was inoculated with a hearty BM-45 strain of yeast to start fermentation. The floating skins were manually punched down twice daily for eleven days. The wine was then allowed to sit with the skins for an additional two days prior to pressing to soften the tannins.

After a gentle pressing, the resulting wine was transferred to French Oak barrels made by Tonnelleries Taransaud and Francois Frere where malolactic fermentation finished. After six rackings and a fining with fresh egg whites, it was bottled in May 2000.

Related information:

Brix at Harvest: 26.1   **Harvest Date**: October 1998
pH: 3.42   **Total Acid**: .84gm/100ml
Bottling Date: 5/2000   **Release Date**: September 2000
Production: 475 cases

W i n e m a k e r ' s   N o t e s : This wine is from one of the premium hillside Napa Valley vineyards, and is aged primarily in French Oak. The '98 shows rich, dark black cherry with hints of vanilla and orange zest in the bouquet. The flavors are blackberry and currant with evident hints of menthol and chocolate. One of our best aging Zinfandels.

1998 Napa Valley Zinfandel Lyons Vineyard

Vineyard: Lyons Vineyard, Napa Valley
Cap and Sylvia Lyons, proprietors
This vineyard is located almost 1400 feet above the Napa Valley on the eastern slope of Vaca range. The sparse producing vines are grafted to the 100-yeard old Hambrecht clone of Zinfandel. The hillside location, sparse production, small berries, and mountain climate adds a tremendous intensity to the wine.
Composition: 100% Zinfandel
Vinification and Maturation: The grapes for this wine were gently crushed directly into the small fermentation bins and cold soaked at 50°F for two days. The fruit was inoculated with a BM-45 strain of yeast. A warm fermentation ensued with temperatures into the low 90's. The must was hand punched down twice daily. After seven more days on the skins, the wine was pressed and racked into 100% new French Coopered American Oak. It was racked four times before being bottled unfined in May of 2000. This is one of four small production lots bottled in a special tall gold painted bottled.

Related information:

Brix at Harvest: 25.4   **Harvest Date**: October 16th, 1998
pH: 3.56   **Total Acid**: .78gm/cl

Bottling Date: May 2000 **Release Date**: July 2000
Production: 240 cases available at the winery only

Winemaker's Notes: This exotic mountain Zinfandel exhibits ripe cherries and blackberries with hints of dark currants, Asian spices and vanilla. It has a good tannin and acid backbone, which makes cellaring for 5-7 years a worthwhile effort. It definitely is a Zinfandel that will roar when you pair it with a Rack of Lamb or grilled Venison.

1998 Alexander Valley Zinfandel Harris-Kratka Vineyard

Vineyard: This 20-acre vineyard in the rolling hills just east of the Russian River is located at the corner of Highway 128 and Chalk Hill Road. The vineyard is planted with 47-year old head-pruned vines, 80% of which are Zinfandel and 20% are Carignane. The vineyard has a red clay loam soil. Production is typically 1.5 to 2 tons per acre.
Composition: 78%% Zinfandel 22% Carignane
Vinification and Maturation: The Zinfandel and Carignane were harvested and vinified separately. The grapes were crushed to small open top fermenters with 20% whole clusters. The must was hand punched down twice daily and allowed an extended post-fermentation maceration for 8 days. Aging occurred in 60 gallon French and American oak barrels with about 40% new oak for 16 months. The Zinfandel and Carignane were blended in April 1999 to create the final version of this wine.

Related information:
Brix at Harvest: 25.8   **Harvest Date**: October 12th, 1998
pH: 3.61   **Total Acid**: .60gm/100ml
Bottling Date: 2/2000   **Release Date**: March 2000
Production: 768 cases

Winemaker's Notes: Fruity and intense, this wine is one of our finest Harris-Kratka Zins to date. It show intense black cherry and ripe plum fruit typical of the Alexander Valley. It is smooth, silky and exotic on the palate with hints of violets, vanilla and currants.

1998 Sonoma Valley Zinfandel

Rosenblum Cellars, Island of Alameda   256

Cullinane Vineyard

Vineyard: This 99 year-old head pruned, dry farmed vineyard is located in the southeastern portion of the town of Sonoma about 3 blocks from the Carneros region. The soil is gravely Sonoma loam with a yield of about 1½ tons per acre. The vineyard is bordered by eucalyptus trees, which may be why this wine exhibits menthol characteristics.

Composition: 100% Zinfandel

Vinification and Maturation: The grapes were hand harvested late in October 1998, and fermented in small vats using our labor intensive hand punch down method with about 15% whole cluster fruit and a native indigenous vineyard yeast. The wine was gently pressed and transferred to new American and two year old French barrels. During aging, it received three rackings before being bottled unfiltered and unfined in May 2000. It is one of four small production wines bottled in a special tall gold painted bottle.

Related information:

Brix at Harvest: 26.8   **Harvest Date**: October 26th, 1998
pH: 3.50   **Total Acid**: .82gm/100ml
Bottling Date: 5/2000   **Release Date**: June 2000
Production: 125 cases available at the winery only

W i n e m a k e r ' s   N o t e s : From this old vineyard comes a wine showing cracked black pepper with black raspberry and hints of mint and oregano. This 1998 is a huge wine that carries its 15.2% alcohol with grace and dignity. It was grapes from this vineyard that founder and winemaker Kent Rosenblum used to produce his first award-winning Zinfandel, way back in 1974.

1998 San Francisco Bay Zinfandel Continente Vineyard

Vineyard: Proprietor, John A. Continente, Oakly, Contra Costa County
This fine, old, head pruned, dry farmed vineyard sits on sandy soil just west of Oakley in the shadow of Mt. Diablo. Contra Costa County is located 40 miles east of San Francisco and 30 miles south of the Napa Valley. The vines which produced these grapes are "new" plantings for the Continente family, being only 50 to 100 years old whereas other parts of the vineyard were planted as far back as 1878.

Composition: 100% Zinfandel

Vinification and Maturation: The grapes for this wine were hand harvested in late October 1998, almost two months later than usual. After being transported to Alameda, they were crushed into a stainless steel fermentation tank and cold soaked for three days. Fermentation was started by gradually raising the tank temperature to 88°F to activate the indigenous yeast. The wine was pumped-over daily to increase the extraction of both color and flavor. The wine was then pressed and aged in American oak 60 gallon barrels and racked four times before being bottled.

Related information:

Brix at Harvest: 26.	**Harvest Date**: October 1998
pH: 3.65	**Total Acid**: .80gm/100ml
Bottling Date: 2/2000	**Release Date**: March 2000
Production: 2130 cases	

Winemaker's Notes: This wine exhibits rich chocolate plum and cherry character in the bouquet with ripe black currants and hints of exotic spices and berries in the flavors. This wine will nicely compliment hearty stews, grilled meats and southwestern cuisine.

1998 San Francisco Bay Zinfandel Contra Costa County

Vineyard: 54% Cutino, 26% Plachon, 16% Pato, and 4% Meadows Vineyards.

Contra Costa County is in the San Francisco Bay appellation and is geographically situated at the inlet to San Francisco Bay, about 30 miles south of the Napa Valley, and slightly northeast of the Livermore Valley. The vineyards we used for this wine contain some vines close to 100 years old, planted in sloping, sand and well-drained soil. The phylloxera louse does not like sand and thus many of these vines survived the epidemic in the early 1900's These vines are all dry farmed, virtually without irrigation.

Composition: 88% Zinfandel 8% Carignane 4% Petite Sirah

Vinification and Maturation: This luscious and elegant example of Contra Costa County Zinfandel is crafted from old vine and head pruned, dry farmed vineyards. This wine is a blend of the above mentioned vineyards which also have old Carignane, Petite Sirah and Mourvèdre vines. Each varietal is vinified separately and then used in the final blend. The grapes were harvested in early September and transported to Alameda for crushing. The crushed fruit and 15% whole clusters received a cold soak at 50°F for three days before being warmed to 70°F where an induced yeast fermentation ensued. The wine was pressed after 3 days extended maceration and aged in 60 gallon oak barrels, half French, half American, with 15% new.

Related information:
Brix at Harvest: Average of 24.4
Harvest Date: Various in early September 1998
pH: Average of 3.62
Total Acid: Average of .68gm/100ml
Bottling Date: 12/99   **Release Date**: January 2000
Production: 2500 cases

W i n e m a k e r ' s   N o t e s : Our 1998 Contra Costa County Zinfandel shows an elegant bouquet of ripe cherries and plums with nuances of strawberry, pepper and vanilla. The flavors are fresh ripe blackberry with essence of Santa Rosa Plums and spicy currants. This is a great wine to accompany pizza, barbecue and pasta of all kinds.

1998 San Francisco Bay Zinfandel Carla's Vineyard

Vineyard: Carla's vineyard, run by the Cutino-Meadows family, is located in Contra Costa County just north of the Antioch bridge. The vines are over 100 years old, head pruned, and dry farmed. The soil is well-drained sand and gravel. These gnarly stumps of vines produce beautiful bunches of fruit with moderately small berries and thick skins, with production a moderate 1.5 tons per acre. This low production adds to the intensity of the wine. The vineyard is named after Carla Meadows who, with her husband Dwight, have led the fight to keep this historic vineyard from becoming a shopping center parking lot.
Composition: 88% Zinfandel 12% Carignane
Vinification and Maturation: These fine looking grapes were harvested in late September 1998. They were crushed into a stainless steel tank and cold soaked at 52°F for 3 days. The temperature was increased to 70°F, and cultured yeast fermentation started and lasted eight days at a peak temperature of 91°F. The grapes were gently pressed and the resulting wine racked into barrels, which were 60% American and 40% French with 30% being new oak.

Related information:
Brix at Harvest: 24.8   **Harvest Date**: September 16th,1998
pH: 3.46   **Total Acid**: .76gm/cl
Bottling Date: 1/2000   **Release Date**: March 2000
Production: 1382 cases

W i n e m a k e r ' s   N o t e s :   This is one of the finest Contra Costa County Zinfandels we have had the privilege of making. It exhibits a huge core of black cherry fruit with elements of bramble, cassis, and vanilla in the bouquet. The flavors are ripe black cherry and blackberry with hints of spicy cranberry and black olive in the flavors. The wine goes very well with spicy red sauced penne pasta, accompanied by melon wrapped in prosciutto.

1998 Napa Valley Zinfandel Ballentine Vineyard

Vineyard: Ballentine, Northern Napa Valley
The Ballentine Family has been actively growing and harvesting grapes in the Napa Valley for well over 100 years. Located in the northern Napa Valley between Highway 29 and the Silverado Trail, this vineyard is a prime example of premium dry farmed old vine Zinfandel. The vines were planted in 1921. The alluvial gravel and volcanic loam soil of the Napa Valley help to make these grapes some of the finest fruit from this area.
Composition: 100% Zinfandel
Vinification and Maturation: Picked at the peak of ripeness, the fruit was quickly crushed into a stainless steel tank. After soaking for four days, cultured yeast was introduced, and the fermentation ensued for seven days, reaching a peak temperature of 88°F. Post fermentation maceration lasted 10 days after which the wine was pressed to tanks to finish malolactic fermentation and settle. The wine was transferred to 20% French and 80% American oak in December 1998.

Related information:
Brix at Harvest: 25.1   **Harvest Date**: October 1998
pH: 3.44   **Total Acid**: .82gm/100ml
Bottling Date: 5/2000   **Release Date**: July 2000
Production: 50 cases

Winemaker's Notes: This wine is a truly fine example of the heights Zinfandel from the Napa Valley can achieve. The wine's appearance is very vibrant, and rich with a nearly opaque garnet red color. The nose soars from the glass, offering sweet wild blackberry pie that is lightly draped in layers of silky brown sugar, cinnamon and cloves. It also exhibits an unctuously textured palate with an explosive cornucopia of blackberries, raspberries and strawberries with a heady finish that lasts into next week.

1998 Sonoma County Zinfandel Russian River Valley, Alegria Vineyard

Vineyard: Alegria, proprietor Bill Nachbaur
Located just south of Healdsburg and west of highway 101, this vineyard has vines ranging in age from 50 to 102 years old. All are head trained with an overhead 2-wire trellis system. Seven separate block of well draining loam and clay soil slopes are all planted to various percentages of Zinfandel, Alicante Bouschet and Petite Sirah.
Composition: 81% Zinfandel 12% Petite Sirah 7% Alicante Bouschet
Vinification and Maturation: Each block of this vineyard is harvested separately as it ripens. The individual lots are kept separate after the hand punch down small vat fermentation. Following the first racking the lots were evaluated for the master blend. The resulting wine was aged in a 70/30 mix of American and French 60-gallon oak barrels with half of the American oak being new.

Related information:
Brix at Harvest: 25.8   **Harvest Date**: October 1998
pH: 3.49   **Total Acid**: .74gm/100ml
Bottling Date: 3/2000   **Release Date**: June 2000
Production: 900 cases

Winemaker's Notes: This wine is a fine example of a harmonious vineyard blend that focuses on richness, color and structure, due primarily to the high percentage of Alicante and Petite Sirah in the vineyard. The wine itself shows intense amounts of chocolate, cassis and blackberry with a slight hint of black olive in the nose. The flavors are blackberry jam, ripe plums, cassis and chocolate with a nuance of mint. We think this wine will be a superb candidate to age in the cellar for 8-10 years.

Rosenblum Cellars also produces Cabernet Sauvignon, Meritage, Mourvèdre, Merlot, Petite Sirah, Port, Carignane, Semillon, Chardonnay, Black Muscat, a Semillion and Chardonnay blend, Gewürztraminer and a Sparkling Gewürztraminer.

Saucelito Canyon

Saucelito Canyon Vineyard
1600 Saucelito Creek Road
Arroyo Grande, CA 93420
nancy@saucelitocanyon.com www.saucelitocanyon.com
office phone and fax: 805-543-2111/ 805-543-0553

We are located in a beautiful and remote canyon in the upper Arroyo Grande Valley, approximately 20 miles inland from San Luis Obispo. The warmer climate, moderated ocean influences, and porous soils combine to create a unique Zinfandel growing region. We have eight acres of Zinfandel, three of which are old vines, and one acre of Cabernet Sauvignon, Merlot, Cabernet Franc combined. Saucelito Canyon is a small family owned winery, making only Zinfandel and Cabernet Sauvignon from estate grown fruit. We share a tasting room with Talley Vineyards in the historic Rincon Adobe.
Tasting room: 3031 Lopez Drive, Arroyo Grande, CA 93429 (805-489-0446)
Owners: Bill and Nancy Greenough   **Winemaker**: Bill Greenough

Winemaking Philosophy:
Our philosophy is one of minimalist winemaking that allows the fruit character to show through our wine. Century-old vines and a unique microclimate combine to create lushly fruited wines with a spicy backbone. We believe great wine starts in the vineyard. We dry farm the vines, regulate the canopy for maximum sun exposure, and thin the crop to ensure the best quality. Our goal is to let the vineyard express itself through our wines.

The Vineyards:
Our vines are recognized as the oldest living Zinfandel vines in San Luis Obispo County, dating back to 1879. Cuttings from the vineyard were collect by the University of California at Davis and planted in their Heritage Zinfandel Vineyard: a vineyard comprised of the oldest Zinfandel clones in California.

Our vineyard consists of eight acres of Zinfandel. The original three acres of vineyards were planted in the late 1870's by homesteader Henry Ditmas. My husband, Bill, purchased the vineyard from Ditmas' granddaughters in 1974 and we bonded the winery in 1982. The old vines still maintain their original head-pruned shape and the vineyard is dry farmed. We use just our own fruit and produce approximately 2000 cases of Zinfandel each year. Saucelito Canyon

Vineyard is located in upper Arroyo Grande Valley of San Luis Obispo County. Temperatures are cooler than Paso Robles but warmer than Edna Valley and lower Arroyo Grande Valley. The maritime influence and porous limestone, sand, and silt-tone soils create an outstanding terroir and microclimate for growing dry-farmed Zinfandel.

The 1998 vintage in Amador was one of the coolest on record - we picked our last grapes the day before Thanksgiving. Because of our organic farming, small crop, and loose bunches, we had none of the bunch rot or unripe grapes that most other areas experienced.

1997 Zinfandel

Vineyard: Vines were first planted on Rancho Saucelito (saucelito is an old place name for little willow that grow naturally in the area) in 1879 by Englishman Henry Ditmas. When Nancy and I bought the property from Ditmas' granddaughters in 1974, the vines were overgrown, having been abandoned for more than three decades. Vine tops were dead, but the roots were still alive, so I picked one shoot from each root crown, then cut off the rest to create a new trunk. Our soils are deep and rich in sand, gravel and limestone. The land is uplifted ocean terrace - now 800 feet above sea level - and fossilized oyster beds come up under the plow.
Composition: 100% Zinfandel
Vinification and Maturation: We try to pick the vineyard three separate times - early, mid, and late to capture the qualities from each stage. We ferment lots separately in open top stainless steel fermenters and barrel age 9 to 10 months in small oak cooperage. New oak is minimal, about 10% a year. We rack infrequently, usually just once to combine lots before bottling.

Related information:
Alcohol: 13.5%   **Release Date**: April, 1999   **Production**: 2727 cases

Winemaker's Notes:
The 1997 Zinfandel gives a velvety mouth feel and boasts aromas of wild strawberries, black cherries, and plums with hints of rose petals and herbal notes of licorice. Toasty oak and traces of vanilla integrate in this deeply layered wine. A very versatile food wine ready to drink now or continue to age for three to seven years.

Sausal Winery

Sausal Winery
7370 Hwy 128
Healdsburg, CA 95448
Phone: 707-433-2285 Fax: 707-433-5136

Now in its third generation of winemaking in Alexander Valley, the Demostene family purchased Sausal Ranch in 1956. At that time the 125 acres were planted to prunes, apples and grapes. However, since grapes and winemaking were most important to Leo Demostene, their father, he began replanting the prunes and apples in the hope of establishing his own winery. Unfortunately he died in 1973 before he achieved his dream. But he did leave behind a set of plans converting the old prune dehydrator to a winery. His four children carried out his dream and Sausal Winery was established and open for business in the fall of 1973. They grew along with the industry and opened a tasting room in the summer of 1986.

Winemaking Philosophy:
The Demostene Family of Sausal Winery would like to introduce you to their quality wines. From the tending of the grapes to the pouring of the wines, we devote our time and energy to blending our history, our expertise and our constant care into wines for your enjoyment.

1998 Zinfandel

Vineyard: Estate grown with 65 year old vine dominating.
Composition: 100% Zinfandel
Maturation: Aged 16 months in French and American oak.

Related information:
 Alcohol: 14.1% **Residual sugar**: Dry
 TA: 0.60 **pH**: 3.70
 Release Date: June, 2000 **Production**: 6000 cases
 Aging potential: 3-5 years

Winemaker's Notes:
Sausal Winery, Alexander Valley   265

Rich, warm and complex. Deep extracted berry-like fruit highlight the wine, while sweet oak and spice add to the character and charm.

1997 Private Reserve Zinfandel

Vineyard: Estate grown produced from vines at least 85 years old.
Composition: 100% Zinfandel
Maturation: Aged 22 months in French oak.

Related information:
Alcohol: 14.1%	**Residual sugar**: Dry
TA: 0.68	**pH**: 3.74
Release Date: February, 2000	**Production**: 2000 cases
Aging potential: 5-8 years	

Winemaker's Notes:
Intense raspberry, oak and ample tannins give testimony to the heritage of 85 to 100 year old vines. Deep and rich, this wine can be enjoyed now or cellared for future appreciation.

1997 Century Vine Zinfandel

Vineyard: Estate grown 120 plus year old vines.
Composition: 100% Zinfandel
Vinification and Maturation: Aged 22 months in French oak.

Related information:
Alcohol: 14.6%	**Residual sugar**: Dry
TA: 0.76	**pH**: 3.60
Release Date: November, 1999	**Production**: 830 cases
Aging potential: 5-8 years	

Winemaker's Notes:
From our oldest vines, well over 100 years, comes this rich velvety Zinfandel. Bursting with berry fruit this wine ends with a soft silky finish. A unique, elegant style Zinfandel.

Sausal Winery also produces Cabernet Sauvignon, Sangiovese and a proprietary white; Sausal Blanc.

Saviez Vineyard

Saviez Vineyard
4060 Silverado Trail
Calistoga, CA 94515
Phone: 707-942-5889 Fax: 707-942-6037
Email: grapeman@juno.com
Website: www.saviezvineyards.com

The Saviez family have continually been growing grapes in the Napa Valley for more than a century. In 1890, Francois Saviez migrated from France and started working for Lilly Coit of San Francisco. "O'Fireball Lilly", as she was known, owned vineyards in the Larkmead area near Calistoga. Francois farmed these vineyards for Mrs. Coit until 1900. At that time he purchased about 200 acres that became what is still the Saviez family farm.

In the 1920's he opened and operated a winery which was destroyed by a forest fire in the 1940's. The wine industry was in recession, so he opted not to rebuild. He continued to farm the vineyards until Cyril, his son, took over. In the 1980's, Paul, Cyril's son, took over management of the farm. Paul is hopeful that his daughter, Monique, will be the fourth generation to continue the family tradition. The renewed interest in fine wine has prompted Paul to revive the family heritage of wine making.

Owner: Paul Saviez   **Winemaker**: Dennis Johns

Winemaking Philosophy:
Saviez Vineyards has always produced great grapes, and when I decided to start producing wine I knew I would need the best wine maker available to complement the fruit. With that in mind Dennis Johns, from White Cottage and St. Clement Vineyards, agreed to work his magic on the Zinfandel. Picking the fruit when it was at its peak, fermenting as only Dennis can do and aging in Hungarian and French Oak barrels has produced what I think is a quality Zinfandel of the highest standard.

The Vineyards:
The Saviez Vineyards are located in the Larkmead area of the Napa Valley. The rocky, alluvial soils and the hot, stressful microclimate of this area produce fruit with very intense flavors that is conducive to the development of great Bordeaux varietals. The vines are derived from 100 year old stock from the Dry Creek area. These clones are known for their smaller bunches, powerful flavors and deep red

color. The farming is tough, but results are unbeatable!! The Larkmead area has a history of producing exceptional Zinfandel. For this reason Zinfandel was chosen for the first release. The fruit used to produce the 1998 vintage was grown within a mile radius of Larkmead Lane in the vineyards either owned or managed by Paul. Control of the vineyards means control over the quality of the fruit, the optimum time for harvest ultimately - great wine.

The Vintage:
The 1998 vintage year was another to remember. The winters of this decade do seem to be going through a cycle of starting later and lasting longer into the spring, as did the 1998 season. The rains lasted until the first week of June causing havoc as far as early shoot Botrytis and mildew, but was kept in check with good farming techniques. We had a seemingly short summer that was very mild except for a couple of weeks in July that were extremely hot. 114 degrees in places causes some burn. The harvest was also interrupted by a mid-season early rain that panicked some into picking too early, but others, I am happy to say, held on to pick at the opportune time. The fruit, despite all the tribulations reached its optimum maturity and was harvested at its peak.

1998 Zinfandel, Napa Valley

Appellation: Napa Valley
Vineyards: Saviez Vineyards and Vukic Vineyard
The Saviez Vineyards are located four miles south of Calistoga on the Silverado Trail at the crossroad "Larkmead Lane." This area is long known for its special microclimate which is ideal for producing very intense flavored grapes. Coupled with its Alluvial soils, the area produces some of the best Zinfandel grapes available. The Vukic Vineyard is also managed by Saviez Vineyard Management and is located adjacent to Saviez Vineyards. The vines and microclimate are the same as ours, but coupled with a slightly heavier based soil that adds a little different texture to the great Zinfandel it produces.
Composition: 100% Zinfandel

Related information:
Alcohol: 15.0%   **Residual sugar**: Dry
Bottling Date: March, 2000
Release Date: July, 2000
Production: 450 cases
Retail: $28.00

Winemaker's Notes:
The '98 has complex aromas of lush berries intermingled with floral bouquets. It exhibits rich, full bodied berry flavors which lend to its jammy characteristics with hints of spice, and the oh-so-familiar peppery finish, which begs to linger on your palate for the sheer excitement and pleasure it brings to ones senses. Backed by lasting silky tannins it provides a remarkably smooth finish. This is truly a unique Zinfandel to be enjoyed now and in years to come.

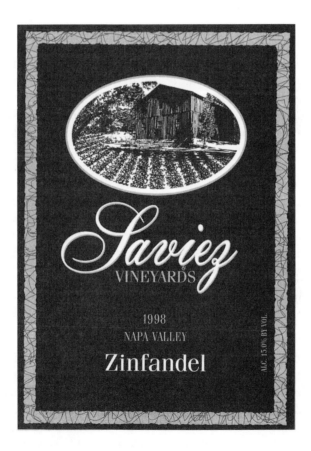

Saxon Brown Wines

Saxon Brown Wines
P.O. Box 1832
Sonoma, CA 95476
Phone: 707-939-9530 Fax: 707-939-9535
Email: jeff@saxonbrown.com

Our family has lived in Sonoma Valley for four generations, and various members have been involved over the years in grape growing and winemaking. I can't remember a time when I didn't want to make wine, and to this day, I can't imagine a time will ever come. Saxon Brown comes as the result of that desire, coupled with close to twenty years of viticultural and winemaking experience with a variety of premium Sonoma Valley wineries. Most notable were my fifteen years with Chateau St. Jean, where I was fortunate to learn from and collaborate with two of the best winemakers in the industry; Richard Arrowood, of Arrowood Winery and Don Van Staaveren, of Artesa Winery. It was during these years that I developed my philosophy of producing small lots of wine from the very best vineyard sources that I could find and to do so without compromise - the best grapes - the gentlest winemaking techniques - and lots of hands-on attention. I believe that winemaking must consume you or the wine will not be worth consuming.

The Vineyard:
Casa Satinamaria Vineyard was planted before the turn of the century on one of the most magnificent sites in all of Sonoma Valley, perfect for wine grapes. Located north-west of the town of Sonoma, on the Valley floor, the vineyard is planted in a classic field blend of 80% Zinfandel, with the balance made up of Petite Sirah, Carignane, Alicante Bouschet and Mataro. The vines are head pruned and dry farmed. Due to the age of the vines and the fact that they have been dry farmed, the yields are low, the clusters are very small and the fruit is very concentrated. In 1997 (a "big year" volume wise) we were able to harvest 0.83 tons per acre, as opposed to the 4-6 tons per acre which is the average for most Zinfandels in Sonoma Valley. We don't get much, but what we get is out of this world!

1998 Zinfandel Casa Santinamaria Vineyards

Appellation: Sonoma Valley
Vinification and Maturation:
I am a traditionalist and a firm believer in picking by taste and appearance - which is to say that I pay attention to the numbers, but am not ruled by them. In 1998 the Brix at harvest was approximately 25 when the fruit was right to pick. The fruit was transported with optimum care in small bins and in the cool of the day to the winery, where it was de-stemmed into an open-top fermenter retaining 95% whole berry. The fruit was cold soaked for 24 hours, and inoculated with a low rate of yeast - this allows the wild yeast to have an effect before the cultured yeast takes over.

After fermentation begins, we punch-down 3 to 4 times daily. The must was pressed at approximately zero Brix and allowed to finish fermentation in a tank. After fermentation was completed, the wine was racked to new, one and two year old French barrels.

Related information:
Alcohol: **Residual sugar**: Dry
Brix at Harvest:25.0   **Bottling Date**: Fall of 2000
Release Date: January 21, 2001 (my wife's birthday)
Production: 130 cases sold through the mailing list
Retail: $38.00

Winemaker's Notes:
WOW! This vineyard always amazes me with it's intense raspberry and spice signature, a true example of terroir. Very intense and concentrated. As I have said before, this wine is not for beginners. Unfortunately the 1998 yield is even smaller than the 1997 - oh well. Nose: Intense bright fruit, raspberry, spice, blackberry, dried berries, black cherry and plum. This wine reminds me of walking through a berry patch. Mouth: Good structure, very complex, almost overwhelming. Great mouth feel.

One of the best wines I have ever made.

Schuetz Oles

Schuetz Oles
P.O. Box 834
St. Helena, CA. 94574
Phone: 707-963-5121 Fax: 707-967-9431

Schuetz Oles is owned and operated by Rick Schuetz, unassuming and philosophical winemaker, and Russ Oles, local color and notorious grape grower. Rick and Russ believe in hands on involvement in every aspect of grape growing, winemaking and marketing.

No tasting room. Limited mail order.

Proprietors: Rick Scheutz and Russ Oles   **Winemaker**: Rick Scheutz
Grape Grower and farmer: Russ Oles

Winemaking Philosophy: They take their wine seriously, not themselves.

1998 Zinfandel Napa Valley Korte Ranch

Appellation: Napa Valley **Vineyard Designation**: Korte Ranch
Composition: 100% Zinfandel from the Korte Ranch.
Vinification: Hand picked. Small open top fermentation eight to ten days. Hand punched down. Fermented with BM-45 yeast and natural flora malolactic fermentation. Racked four times for clarification. No fining or filtering.
Maturation: 16 months in 50% French and 50% American oak.

Related information:
Alcohol: 14.8%   **Residual sugar**: 0.10%
Brix at Harvest: 24.5
Harvest Date: 10/02/98, 10/17/98 and 10/19/98
Bottling Date: 5/01/00   **Release Date**: 06/01/00
Production: 1,500 cases produced
Retail: $18.00

Winemaker's Notes: The Zinfandel at Korte Ranch was planted about 100 years ago during the administration of Theodore Roosevelt. It has

Schuetz Oles, Napa   272

lived through the Bolshevik revolution, the Red Scare and the dissolution of the U.S.S.R.. These vines have coexisted with men on the moon and survived the invention of LSD, MTV, Velveeta and Cool Whip. This 40 acre parcel just north of St. Helena is owned by Hal Pagendarm whose grandfather owned and operated the Korte Ranch Winery before Prohibition. Up on the hill are the crumbled remains of a stone cellar and house where the original Korte Winery was located. Periodic Elvis sightings are reported by the locals. Ripe wild berry and dried cherry flavors. Full body, long spicy finish, going to licorice and vanilla.

Food pairing: We recommend that people drink wine with every meal - regardless - use your own taste buds - not someone else's.

1998 Esther's Reserve Korte Ranch Zinfandel

Vineyard: 100% Zinfandel from the Korte Ranch
Vinification: Fermented with BM-45 yeast with natural flora malolactic fermentation. Racked for clarification.
Maturation: 16 months in 50% French and 50% American Oak.

Related information:
Alcohol: 15.03%	**TA**: 0.87
pH: 3.51	**Harvest Date**: October 2, 17 & 19th, 1998
Production: 1500 cases	**Retail**: $23.00

Winemaker's Notes: This special barrel select Zinfandel was named in honor of Esther Guifoyle who has lived on the Korte Ranch for over 38 years with her collection of goats, burros, dogs, boars, cats and birds.

1998 Vesuvium

Vineyard: 75% Zinfandel from the Korte Ranch and 25% Cabernet Sauvignon from Rutherford.
Vinification: Fermented with BM-45 yeast with natural flora malolactic fermentation. Racked for clarification.
Maturation: 16 months in 50% French and 50% American Oak.

Related information:
Alcohol: 14.7%	**TA**: 0.65
pH: 3.84	**Production**: 374 cases

Retail: $25.00

Winemaker's Notes: This explosive blend of Zinfandel and Cabernet Sauvignon from the volcanic soils of the Napa Valley erupts with earth-shaking flavors. Lush ripe blackberry, cherry and chocolate flavors. Full body, long spicy finish, going to george and licorice. Zin front, cab back, with the complexity of both.

1998 So Zin

Vineyard: A blend of Korte Ranch and Mendocino Zinfandels, Petite Sirah and Syrah
Vinification: Fermented with BM-45 yeast with natural flora malolactic fermentation. Racked for clarification, no filtering or fining agents.
Maturation: 14 months in 25% French and 75% American Oak.

Related information:
Alcohol: 14.3%	**TA**: 0.62
pH: 3.70	**Production**: 1500 cases
Retail: $12.00	

Winemaker's Notes: Prior to the forming of the League of Nations, local growers blended Zinfandel with Petite Sirah for themselves. In this tradition, we have blended Korte Ranch Zinfandel and Mendocino Zinfandel with Petite Sirah and Syrah. In honor of these old farmers we toast, "This Zest Zinfandel was Zermeoniously Zapped from a Zygot to be Zipped by Zealous Zymurgists. Drink Zinfandel and Zerp". This 1998 So Zin exhibits cherry, blackberry jam and vanilla nuances. Medium bodied, with a fruity, soft finish.

Schuetz Oles also produce a Chappell Vineyard Chardonnay, Rattlesnake Acres Petite Sirah and a Napa Valley Port made from Zinfandel and Petite Sirah.

Sierra Starr Vineyard and Winery

Sierra Starr Vineyard and Winery
11179 Gibson Drive
Grass Valley, CA 95945
Phone: 530-477-8277 Fax: 530-265-2373
Email: sierrastarr@leo.net
Website: www.sierrastarrwine.com

Sierra Starr Vineyards and Winery is located in the foothills of the Sierra Nevada at 2400 ft. Phil and Anne Starr bought an existing vineyard in 1995 that was planted in 1982. They also lease a vineyard planted in 1979. A small family-run winery, they do all of the work with occasional help from their college sons, Jack and Eric when home on breaks. So, Phil is the winemaker and grape grower and Anne is the major gopher. Their first vintage was 1996 with first release in September 1997.

Tasting room is open Saturday and Sunday 12-5 and weekdays by appointment only. Tours available

Owners: Phil and Anne Starr

Winemaking Philosophy:
We believe that a t least 80% of a really good wine is made in our vineyard and finished in the winery. All of our wines are made from Estate grown grapes. We also believe that Zinfandel's great flavors come from ripe grapes above 24 Brix. All of our red wines are fermented in half ton, open top fermenters with hand punch down followed by minimal processing with the bottling of the wines unfiltered and unfined except in rare instances. We use a combination of top quality French and American oak cooperage with only the right amount of new wood so that the wonderful fruit flavors are not over oaked!

1997 Old Clone Zinfandel

Appellation: Sierra Foothills
Vineyard: Stellar Vineyard, Estate
Composition: 100% Zinfandel
Vinification and Maturation: Small, open-top fermenters, hand punch down. Aged twenty months in our cave, bottled unfined.

Related information:
Sierra Starr Vineyard and Winery, Sierra Foothills

Alcohol: 15.5%
Brix: 28 after soak
Harvest Date: October 1997
Bottling Date: June 1999
Release Date: November 1999
The 1998 Vintage will be released in October 2000
Production: 210 cases
Retail: $16.00 Our wines are all priced to be a great value

Winemaker's Notes:
Look for a big jammy flavor with plums, raisins and black cherries.

Sierra Starr Vineyard and Winery also produces Cabernet Sauvignon, Merlot, Chardonnay and Sauvignon Blanc

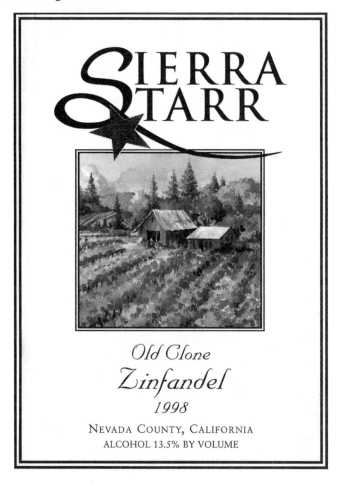

Old Clone
Zinfandel
1998
NEVADA COUNTY, CALIFORNIA
ALCOHOL 13.5% BY VOLUME

Sierra Vista Winery

Sierra Vista Winery
4560 Cabernet Way
Placerville, CA 95667
Phone: 916-622-7221

Sierra Vista Vineyards and Winery is located in the Sierra Foothills area of Northern California, fifty miles west of Lake Tahoe in the Sierra Nevada Mountain at an elevation of 2,800 feet. Forty three acres of vineyard were planted on the installment plan from 1972 until 1982 and are the sole source of grapes used to produce Sierra Vista Wines. The first wines and the first Zinfandel were produced in 1977.

Tasting Room and picnic area open daily 10:00 am to 5:00 pm except major holidays. Mail order. Wine Club. Tours by appointment.

Proprietor: John and Barbara MacCready   **Winemaker**: John MacCready
Assistant Winemaker: Michelle MacCready

Winemaking Philosophy:

The wine says it all. Wines are grown in the vineyard and handled as gently as possible in the winery to preserve the vineyard character. The cool weather in El Dorado where the temperature rarely exceeds 95 degrees make Zinfandel growing a real challenge. Zinfandel needs heat and plenty of air and light to produce fine wine grapes. Our practice of removing suckers before bloom, leaf pulling and hedging after bloom and cluster thinning at verasion is necessary to produce the type of wine I am happy with.

1998 Estate Bottled Zinfandel Herbert Vineyard

The Vineyard : Herbert Vineyard, El Dorado County.

The Herbert Zinfandel vineyard, farmed by Sierra Vista, is on a steep hillside of decomposed granite at 2400 feet elevation and has an old Zinfandel clone with small bunches and small berries. The 1998 vintage of Herbert Zinfandel is bigger than this wine has been in several years. The wine is concentrated with briar, raspberry fruit and highly-extracted black pepper with a long, lingering finish. The vineyard blocks were planted from 1972 until 1976 and are head-trained in the old fashioned style.

The Vintage: The 1998 harvest was a long, drawn out affair starting October 1st with the Herbert vineyard Zinfandel harvested ahead of the Chardonnay and Sauvignon Blanc. Because the crop was short our biggest problem was waiting for the acid to come down so we could harvest. Thus there was good long hang time on the vine which produced an intense balanced wine.

Composition: 100% Zinfandel

Vinification: Once the grapes were hand harvested and hand sorted they were partially crushed and destemmed. Then primary and malolactic took place in stainless tanks that had to heated because of the cold weather at our 2800 foot elevation. Egg white fining and filtering gives an elegant finish to this Zinfandel!.

Maturation: Barrel aged for mostly French oak with a mixture of small American oak, French coopered barrels with medium toast.

Related information:

Alcohol: 14.2%	**Residual sugar**: Less than 0.1%
Brix at Harvest: 24 to 24.8	**Harvest Date**: 10/01/98 to 10/04/98
Bottling Date: 5/9/2000	**Release Date**: Fall

Winemaker's Notes:

Although typically the Herbert Zinfandel is a little lighter in body, just as intense in flavor but not as peppery as the Reeves Zinfandel, this year was an exception to the rule. 1998 gave us the blackberry and spice character and is a big, full-bodied, peppery Zinfandel. This wine has a history of being delicious when young but improving greatly for 5 years. Because of the change in character it has not developed as rapidly as the Reeves so I suspect it will age longer than is normal for the Herbert Zin. So enjoy now or hold for that wonderful treat that will be yours in 3 to 7 years.

Food pairing: Sierra Vista Zinfandel is great with Italian dishes, tomato based sauces and all red meats. It will enhance most ethnic cuisine including Mediterranean, Cajun, Spanish, Mexican and South American. A personal favorite is young, zesty Zinfandel with enchiladas.

1998 Estate Bottled Zinfandel Reeves Vineyard

The Vineyard: Reeves Vineyard, El Dorado County.

While the Herbert Zinfandel was the first picked, the Reeves Zinfandel was one of the last picked with harvest going well into November. Because of the cool weather that time of year, hang time was extended and our main worry was the approach of the rainy season. We were lucky and beat the rain with our entire harvest as well as the Zinfandel.

Composition: 100% Zinfandel

Vinification: The Reeves Vineyard Zinfandel was hand picked, hand sorted, and fermented in stainless steel tanks using pumpovers with malolactic fermentation being completed by the end of primary fermentation. The wine is lightly egg white fined and filtered prior to bottling.

Maturation: Barrel aged in small French coopered American oak barrel 2 to 4 years old.

Winemaker's Notes:

Reeves Zinfandel is a Zin lovers Zinfandel with full body, intense flavors and a very blackberry spicy long finish. This wine has a history of being delicious when young, will be really good in 3 to 4 years and will rival most Cabernets in 5 to 7 years. So enjoy now or hold for that wonderful treat that will be yours in 3 to 7 years.

Food pairing: This robust Zinfandel begs for spicy barbecue dished, Italian red sauces, hot Mexican entrees and even hot Thai dished. Sierra Vista Reeves Vineyard Zinfandel is great with lamb shanks or chops, beef bourgogne, spicy frajitas, or zesty sausages. The more elegant style for this vintage lends itself to herb chicken and chicken cacciatore. This wine is an outstanding choice for pizza, spaghetti and lasagna.

Sierra Vista Vineyards also produces Syrah, Viognier, a Rhone style blend called Fleur de Montagne, Lynelle, a Provence style blend named for our daughters; Lynette and Michelle, Cabernet Sauvignon, Merlot, Fumé Blanc and Chardonnay.

Single Leaf Vineyards

Single Leaf Vineyards
7480 Fairplay Road
Fair Play, CA 95684
Phone: 530-620-3545 Fax: 530-620-7214
Email: singleleaf1@aol.com

Single Leaf is located in the beautiful rolling hills of El Dorado in the Sierra Foothills. The vineyards and winery reside at 2400' in elevation with exposures to the north, east, and west. Owners Pam and Scott Miller purchased the property, which included nine acres of Zinfandel, in 1988. Over the succeeding years and with the able assistance of their three children, Heather, Amber and Adam, they planted another 6 ½ acres of grapes. In 1993, the winery was constructed and opened to the public for tasting.

Owners: Pam and Scott Miller **Winemaker**: Scott Miller

Winemaking Philosophy:

The grapes know best! Our founding philosophy has always been to give the vineyards the best care possible, ripen the grapes fully and produce wine by very traditional means. We seek maximum Zinfandel fruit extraction from grapes grown on head-trained, spur-pruned and dry farmed vines. We let fruit flavors dictate picking times with Brix at harvest typically 24.4 and 25.5 sugar content. Our objective is to create powerful, agreeable Zins.

1998 Single Leaf Zinfandel

Appellation: El Dorado

Vineyard: Estate vineyards, 6x12 spacing, head trained, spur-pruned and dry farmed. Production is maintained at 2.5 tons per acre for the Reserve Estate Zin and 3.5 tons per acre for the Estate Zin. Soils are Holland series sandy-loam of granite derivation with topsoil depth between 5 and 8 feet.

Composition: 100% Zinfandel

Vinification and Maturation: Overall production is generally composed of 6 to 8 lots differentiated by vineyard exposure and yeast strain used to ferment. Open top fermenters are pumped over or punched down twice daily during the 8-17 day fermentation period (depending upon yeast strain). All lots are fermented dry then pressed to settling tanks for 48 hours and malolactic is induced. Both French and American Oak barrels are employed for aging, averaging 20% new wood, for the next 20 months. The wines are racked every 4 months. Lot blending is done 6 months prior to a light filtration and bottling.

Related information:
Alcohol: 14.1%
Brix: 24.4
Residual sugar: Dry
Harvest Date: October 22 - 31, 1998
Bottling Date: May 20, 2000
Release Date: August 2000
Production: 650 cases
Retail: $11.00

Winemaker's Notes:

The 1998 growing season was definitely a challenge. Cold and rain during late May and June slowed growth and extended the season. Thankfully, a warm fall helped out and the grapes hit stride in late October with great acid/pH. The resulting wine has very full mouthed fruit flavor and the prospect of 4-6 years of aging potential. In color and texture, the wine is most similar to the Single Leaf 1990 and 1993 vintages, although the flavors compare best with our 1994.

Cellar Notes:

1990 Zinfandel: Smooth and still holding fruit. Acids and tannins have leveled off.
1993 Zinfandel: Peaked now but still holding tannin and fruit.
1994 Zinfandel: Rich and smooth, moderate tannin, excellent acid.
1995 Zinfandel: The year of serious black pepper. Great with food but still has 2-4 years left in this wine.
1996 Zinfandel: Lighter than 95' but a more supple mouth feel. Great balance with 3-4 tears of aging potential.
1997 Zinfandel: A blockbuster wine. Deep, rich wine that will age for many years.

Single Leaf also produces Chardonnay, Zinfandel Reserve, Cabernet Sauvignon and Port.

Sky Vineyards

Sky Vineyards
3030 Cavedale Road
Glen Ellen, CA 95442
Phone: 707-935-1391

When Lore Olds, founder of Sky Vineyards, began searching for the ideal environment to grow premium quality Zinfandel, he had several criteria in mind. These criteria included red soils, eastern exposure, free draining soils, constant air movement, steep hillsides and at least 2000 foot elevation. It was 1972 when Lore found his ideal site, nestled near the top of Mt. Veeder, astride the Napa and Sonoma county borders. This quiet, beautiful location provided everything that Lore had been searching for at an altitude of just over 2100 feet. Lore purchased the property the following year and named it Sky Vineyards. He named it Sky due to the fact that the sky is what dominates in this wild chaparral country.

The first Zinfandel vines were planted at Sky in 1974. This grew to around 12 acres in 1979 when we became Bonded Winery #4934, just prior to our first vintage. Some of the vineyards have just been replanted. Sky Vineyards currently has around 14 acres of Zinfandel planted and has plans for two acres of Syrah.

The timing of harvest is determined by carefully monitoring the flavor of the grapes in association with total acid and pH levels. While Lore also monitors the sugar levels, they have little importance in the decision of when to harvest. Thus the alcohol levels of the wine tend to vary year to year. Harvest often commences in early September, but on one occasion it didn't begin until October. All grapes are carefully hand harvested in one ton lots and then transported to the winery which is nestled in the valley below the vineyard.

Each one ton lot is fermented separately using Louis Pasture Red yeast. Each ton is hand punched and then pressed using an early 1900's basket press into tanks where the wine remains until spring. It is then transferred into French oak barrels where it remains for 12 to 14 months undergoing malolactic fermentation. The wines are hand bottled, hand corked, stored for one year at the winery, hand labeled and then cellared until sold.

The wines are typically described as having intense fruit and black pepper, along with black cherry, and violet flavors. The mouth is very round and long lasting full, of tannin with a strong acid balance. The wines tend to reach their peak five to ten years following the vintage date.

Each year a new label and cork are produced by Lore uniquely for each wine. A different label helps to demonstrate that each vintage is different and emphasize the creative and artistic talents that went into it's production.

Lore began making wine at Hillcrest Vineyards, Oregon in 1971 and has worked for several Napa/Sonoma vineyards and wineries. He is the winemaker, viticulturist, artist, owner, and a very proud father.

Winemaker's Philosophy:
Vintner Lore Olds believes that Sky Vineyards is the perfect place to grow Zinfandel. He views the grape and his wines as an artistic medium to examine the natural history of the soils and skies (terroir) of Mt. Veeder. He planted all the vineyards himself in the mid-70's, and all the winemaking is done by hand.

1996 Sky Zinfandel

Appellation: Napa Valley
Vineyard: Sky Vineyard, Mt. Veeder.
Highest winery on Mt. Veeder at the headwaters of Devil's Canyon, just on the east (sunrise) side of the ridge separating Napa and Sonoma counties at the southern end of the Mayacamas mountain range. The vineyards at the top of the ridge get early morning sun, followed by oblique sun in the afternoon as the orb arcs to the west. Temperatures are 5 to 10 degrees cooler than the valley in summer, and 25 to 30 degrees cooler in the winter with heavy frost and snow not unusual. Despite a generally late spring bud break (a result of cooler temperatures at the high altitude), the grapes ripened early (usually by early September) thanks to the long days of summer sun.

Dry grown, bush vines, no fertilizers used, the vines are on average 21 to 24 years old and produce 2.5 tons per acre. Original vines are planted on AXR rootstock and a few St. George. The vineyard is being replanted due to phylloxera; the new vines are on 110R and will begin producing in 2004.

Composition: 100% Zinfandel

Vinification and Maturation: The pH is monitored until it reaches about 3.2 and then the flavor and visual aspects are monitored until the grapes are determined to be ripe. The grapes are generally picked when there is a slight wrinkling to the skins. The fruit is picked very carefully and sorted twice to ensure no "mog" (material other than grape) and no damaged fruit gets into the must.

The grapes are crushed directly into one-ton fermenters, which are hand punched for 7 to 10 days. Then the wine is pressed using an early 1900's basket press into tanks where it remains until spring. It then is transferred into French oak barrels (usually between 10 and 20% new) for one year. The wines are hand bottled, hand corked, and stored for one year at the winery, hand labeled and then cellared until sold.

Related information:
Alcohol: 13.1%
Total Acidity: 0.57g/100ml
pH: 3.59
Production: 1550 cases
Harvested by hand: Sept 1 -21

Winemaker's Notes:
Zinfandel is a very transparent grape; it shows everything you do to it, good or bad. Up here in the boonies, it particularly shows 'garigue' – a southern Rhône descriptor for flavors of brush and wild herbs. Not surprising, since the Sky vineyards are in chaparral county, not among the big firs and oak trees of the lower elevations on Mt. Veeder. Zin is the perfect grape to show the wildness of the Mt. Veeder terroir.

Sobon Estate
Shenandoah Vineyards

Sobon Estate and Shenandoah Vineyards
12300 Steiner Road
Plymouth, CA 95669
Phone: 209-245-4455 Fax: 209-245-5156
Website: www.sobonwine.com

Shenandoah Vineyards was founded in 1977 by Shirley and Leon Sobon. They moved to the old Steiner Ranch with their six children, planted a vineyard, and converted the old stone garage to a winery. All of the children helped to build the operation, and two sons and a son-in-law now work full time along with Leon and Shirley in the business.
Open daily 10:00 a.m. to 5:00 p.m.
Owners: Sobon Family Trust   **Winemaker**: Leon Sobon

Winemaking Philosophy:
Shenandoah's philosophy is to produce high quality wines at affordable prices. The Sobon family has made a commitment to being good stewards of the land, and all of our vineyards are farmed organically.

The Vineyards:
In 1989, our family made the decision to start farming organically. After a winter of much research, the following spring we saw the start of our new endeavor. We have never stopped learning. In 1994 we registered with the County Agriculture Commissioner and the State of California as organic growers of grape and walnuts. We now farm 125 acres without herbicides, pesticides, or chemical fertilizers.

1998 Zinfandel Special Reserve

Vineyard: Twin Rivers
Composition: 90% Zinfandel 4% Mourvèdre 6% Grenache
Vinification and Maturation: Lots were fermented separately. Peak fermentation of 84° Blended after fermentation. Aged 10 months in small barrels.
Related information:

Alcohol: 13.4%
Harvest Date: October 17th, 1998
Bottling Date: September 7th, 1999
Release Date: November 1st, 1999
Production: 3300 cases

Winemaker's Notes:
This lovely wine is a rich example of the Amador Zinfandel style. The aromas are reminiscent of ripe blackberries, cherries, and cocoa. The wine is full bodied and only moderately tannic. It is beautifully structured with strong fruit and high acid necessary for long aging. The flavors are rich and complex and the finish very long and satisfying.

1998 Zinfandel Fiddletown

Vineyard: Lubenko Vineyards
Head trained vines planted about 1910 at the 1900 foot elevation in Amador County
Composition: 100% Zinfandel
Vinification and Maturation: Aged in French and American oak.

Related information:
Alcohol: 14.9%
Bottling Date: March 31st, 2000
Production: 784 cases

Winemaker's Notes: This is a stunning multi-faceted wine. It has very intriguing aromas of spice, cherry, and concentrated fruit: restrained, but very complex. The flavors are also very complex with a toasty, dusty-cherry component so typical of the Fiddletown area. The finish is long and lingering, and invites another taste.

1998 Zinfandel Rocky Top

Vineyard: Rocky Top Vineyards
Composition: 97% Zinfandel 3% Carignane
Vinification and Maturation: Aged in French and American oak.

Related information:
Alcohol: 14.6%

Sobon Estate, Shenandoah Vineyards, Amador County   286

Bottling Date: March 30th, 2000

Production: 1700 cases

W i n e m a k e r ' s   N o t e s : This is an outstanding Zinfandel of immense proportions. The very floral, vanilla and raspberry jam aromas are very forward and inviting. The flavor match the aromas with a rich juicy component. The finish is long and velvety-rich.

1998 Zinfandel Cougar Hill

Vineyard: Cougar Hill Vineyards

Composition: 100% Zinfandel

Vinification and Maturation: Aged in French and American oak.

Related information:

Alcohol: 14.9%

Bottling Date: March 31st, 2000

Production: 374 cases

W i n e m a k e r ' s   N o t e s : This is a very stylish and likable wine. It has very forward aromas of licorice and mint chocolate. The flavors are rich, full, and reminiscent of ripe, dusty blackberries and cranberry juice. The finish is crisp and lingering.

1998 Primitivo

Appellation: Shenandoah Valley

Composition: 100% Primitivo

Vinification and Maturation: 11 days, hand punched, maximum temperature 81°, Pasteur Red yeast. Aged 14 months in older French cooperage.

Related information:

Alcohol: 14.1%

Bottling Date: January, 2000

Release Date: February, 2000

Production: 422 cases

W i n e m a k e r ' s   N o t e s :

The Primitivo grape has been proven to be a genetic match to Zinfandel. It has been classified as a separate clone of Zinfandel; by U.C. Davis; and a separate variety by the BATF. The vines and grape clusters are morphologically different than Zinfandel, and can easily be distinguished in the vineyard. The grape, and resulting wine shows many similarities to Zinfandel, and many differences, with enough subtle complexities to warrant this varietal bottling.

It has a dark red-purple hue, and is very rich, but soft. It has aromas and flavors reminiscent of blackberry pie and vanilla ice cream, and a long lingering aftertaste. Made entirely from our organically grown grapes.

St. Amant Winery

St. Amant Winery
One Winemaster Way
Lodi, CA 95240
Phone: 209-367-0646   Fax: 209-477-3066
Email: steamant@aol.com

St. Amant Winery is a small family owned winery that prides itself in producing hand-crafted wines of superb quality. The grapes for all our wines are carefully grown and selected for producing fruit of top quality. Vinification follows old world techniques designed for maximum expression of the terroir of each of the vineyard sites. As with all St. Amant Wines production and availability is very limited.

Proprietors: Tim and Barbara Spencer
Winemaker: Tim Spencer

1998 Mohr-Fry Ranch Old Vine Lodi Zinfandel

Appellation: Lodi
Vineyard Designation: Mohr-Fry Ranch
The Mohr-Fry Ranch Zinfandel comes from an old head pruned Zinfandel vineyard planted in 1944 in the Lodi appellation of Northern California. Mohr-Fry Ranches is a family owned and operated business that has been involved in diversified farming operations since 1855. The grapes for this wine came from a 7.5 acre block planted on their West Lane Ranch, and is head pruned in the typical fashion for that time period. Yields are usually very low.

Vinification and Maturation: Picked ripe and hand sorted, the grapes were fermented in open top fermenters utilizing punch down and pump over for 10 days. Pressed at 4.5 Brix, the wine was racked to 60 gallon American oak barrels, 25% new, where primary and malolactic fermentation continued for two weeks. Ten months of barrel aging complemented the rich Zin fruit.

Related information:

Alcohol: 14.5%   **Residual sugar**: Dry
Brix at Harvest:24.8   **Harvest Date**: 9/29/98
Bottling Date: 9/17/99   **Release Date**: Winter 1999
Production: 475 cases sold at the winery and limited distribution.
Retail: $14.00

Winemaker's Notes:

The 1998 Mohr-Fry Zinfandel is our third vintage working with this particular vineyard and we feel it is our best Mohr-Fry Zinfandel to date. More refined than the 1996 and with more depth and complexity than the 1997, the fruit was picked at the optimum maturity, expressing rich Zinfandel fruit character. Ripe berry aromas complement the full body, with spicy but soft tannins, and layers of fruit lingering on the palate. A very distinctive vineyard that stands out in the array of old vine Zinfandels.

Food pairing: The perfect complement to 'haute Lodi Cuisine'…barbecued ribs, tri-tip and burgers, a hearty pot of chili beans and a stick of French bread are an unforgettable experience with St. Amant Mohr-Fry Zinfandel.

St. Amant Winery also produces Barbera, Roussanne, Viognier, Syrah and grows seven different Portuguese varieties that are used in producing Port Wines and Portuguese table wines.

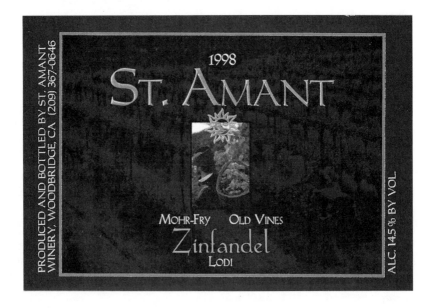

St. Amant Winery, Lodi

Sunset Cellars

Sunset Cellars
405 Alberto Way Suite 7
Los Gatos, CA 95032-5406
Phone: 408-356-2545 Fax: 408-356-2439
Email: sales@sunsetcellars.com
No tasting room. Mailing list

Sunset Cellars was founded in 1997 to produce hand-crafted wines. It emphasizes Zinfandel and other varieties traditionally planted with it, such as Petite Sirah and Carignane. Although the business office is in Los Gatos, Sunset Cellars' wine is made at Rosenblum Cellars' facility in Alameda.
Owners: Douglas and Katsuko Sparks, Richard Kading, Robert Stillman, Doris Boesch and Thomas Lifson
Winemaker: Douglas Sparks

Winemaking Philosophy:

Sunset Cellars' winemaking objective is to express and refine the flavors and characteristics created in the vineyard. Wines are fermented in half-ton containers and punched down by hand. Racking is as gentle as possible, and aging features the restrained use of new oak. The goal is to develop supple wines with ample fruit that reflect the different flavors Zinfandel produces in different regions and vineyards.

1998 Dry Creek Zinfandel

Appellation: Dry Creek Valley
Vineyards: Grapes came from two locations on the east side of Dry Creek Valley, from well established vines up to 80 years old.
Composition: 88% Zinfandel, 10% Carignane, 2% Petite Sirah.
Vinification and Maturation: Grapes were hand sorted at the crusher and fermented in half-ton containers for 11 days. Underwent malolactic fermentation. Aged in French oak barrels and racked by hand as needed. Unfined and unfiltered.

Related information:
Alcohol: 13.9%   **Residual sugar**: Dry
Brix at Harvest: 23.2   **Harvest Date**: 10/02/98

Release Date: Summer, 2000
Production: 325 cases, sold directly and through selected retailers
Retail: $23.00

1998 Russian River Zinfandel

Appellation: Russian River Valley
Vintage: A slow growing season led to a late harvest on Halloween. Yield was low, and ripe flavors developed only as sugar rose to high levels, yet remained balanced by good acidity. The late harvest and a cold winter meant long yeast and malolactic fermentation.
Vineyards: Grapes were from a small vineyard at the top of a ridgeline, over 2000 feet in elevation.
Composition: 85% Zinfandel, 15% Carignane
Vinification and Maturation: After hand sorting and crushing, the grapes were fermented in half-ton containers for 16 days. Underwent malolactic fermentation which completed in spring of 99. Aged in French oak barrels and racked by hand as needed. Unfined and unfiltered.

Related information:
Alcohol: 16.3%	**Residual sugar**: Dry
Brix at Harvest: 28.4	**Harvest Date**: 10/31/98

Release Date: Summer, 2000
Production: 205 cases, sold directly and through selected retailers
Retail: $18.00

1998 Suisun Valley Zinfandel

Appellation: Suisun Valley
Vintage: The harvest was late in 1998. The vines had been stressed the previous year, and yields were kept low. A long hang time developed complex grape flavors, with good acidity.
Vineyards: This was a test lot from a small hillside vineyard near the Napa County line in a promising area for Zinfandel and Rhône style grapes. As a result, the volume rose significantly in 1999, and will increase again in 2000.

Composition: 100% Zinfandel
Vinification and Maturation: Grapes were hand sorted and fermented in half-ton containers for 9 days. With a cold winter, malolactic

fermentation did not complete until spring of 99. Aged in French oak barrels and racked by hand as needed. Unfined and unfiltered.

Related information:
Alcohol: 14.5%   **Residual sugar**: Dry
Brix at Harvest: 24.8   **Harvest Date**: 10/31/98
Release Date: Summer, 2000
Production: 65 cases, sold directly and through selected retailers
Retail: $12.50

Winemaker's Notes:
These three Zinfandels are true to their regions and growing conditions, yet quite different, reflecting the variability of the Zinfandel grape. The 1998 Dry Creek Zinfandel has forward floral and crushed berry fruit on the nose, rounded, lush raspberry and earthy, spicy notes with a subtly tannic finish; a complex and elegant wine. The 1998 Russian River Valley Zinfandel has densely packed fruit aromas, full bodied blueberry, plum and dried fruit flavors, with peppery notes and great length; a powerful yet balanced wine. The 1998 Suisun Valley Zinfandel has bright earth and cherry flavors with a touch of chocolate, and a pleasant follow-through to a clean, long, fruity finish.

Sunset Cellars also produces Petite Sirah, Carignane and Barbera.

Taft Street Winery

Taft Street Winery
2030 Barlow Lane
Sebastopol, CA 95472
Phone: 707-823-2049 Fax: 707-823-8622

      Started by a group of friends and the family of winemaker John Tierney, Taft Street has been producing Sonoma County wines since 1982. Current production is at 35,000 cases.

Tasting Room open Monday through Friday, 11:00 a.m. to 4:00 p.m. Saturday, Sunday and Holidays, 11:00 a.m. to 4:30 p.m. Mail order.

Proprietor: Taft Street Inc. **Winemaker**: John Tierney
Associate Winemakers: John Giannini and Joe Freeman

Winemaking Philosophy: Careful vineyard selection in Russian River, Dry Creek and Alexander Valleys set the standard for quality and value in Taft Street wines.

1998 Sonoma County Zinfandel

Appellation: Dry Creek Valley
Composition: 100% Zinfandel
Vinification: Stainless steel fermentation, pressed dry.
Maturation: Aged in American and French oak for 12 months.

Related information:

Alcohol: 14.2%	**Residual sugar**: Dry
Brix at Harvest: 24.6	**Harvest Date**: September 1998
Bottling Date: 3/29/00	**Release Date**: September 1, 2000

Production: 123 cases sold at winery and distributed.
Retail: $18.00

Winemaker's Notes: A forceful showing of Zinfandel from California's finest Zin soils. From grapes grown in Dry Creek Valley, the wine has an impressive combination of plum, berry and spice. A strong fruit core allows for early tasting, but will reward five plus years of aging.

Taft Street also produces Chardonnay, Sauvignon Blanc, Merlot and Cabernet Sauvignon.

Tarius Wines

Tarius Wines
908 Airway Court #C
Santa Rosa, CA 95403
Phone: 925-838-8957 Fax: 925-838-8958
Email: nora@tariuswines.com
Website: www.tariuswines.com

Tarius is the second half of Sagittarius, sign of The Archer and the sign under which founding partners, Scott Nelson and Tim Olson, were born. Their homemade wines were first crafted under the label Domaine Archer and Archer Cellars but when the two astrology and astronomy enthusiasts decided to make wine commercially it seemed the name Tarius was written in the stars.
Owner and Winemaker: Tim Olson

Winemaking Philosophy:
Located in the Russian River Valley of Sonoma County, Tarius Wines specializes in small lot, hand-crafted Pinot Noirs and Zinfandels. The focus is on terroir and each wine produced is an expression of the vintage and the appellation or vineyard it comes from. The style is Burgundian; hand-crafted, non-interventionist, gravity fed, unfined and unfiltered.

The Vintage:
The 1998 vintage was a return to somewhat normal yields, if not slightly below normal. The vintage got off to a good start with bud break on time in March. Flowering in May saw winds and rain resulting in some shatter during fruit set. This lowered crop yields and raised the prospect of the fruit reaching maturity sooner than normal.

After a warm July, August turned cool, and come September, rain became a factor. Throughout harvest, cool weather and intermittent rain forced winemakers to decide between picking under-ripe grapes before threatened rains or waiting it out, risking ripe, but ruined grapes. We chose to be patient and wait for ripeness. Those that waited it out were rewarded with concentration and quality.

1998 Zinfandel, Mendocino

Appellation: Mendocino
Vineyards: Mendocino is a beautiful county with two distinct grape-growing regions. The cooler, inland Anderson Valley is more suited to Pinot Noir and Chardonnay while the Ukiah and Hopland side of the county is more suited to Syrah and Zinfandel. The heart of this wine comes from two vineyards just outside Ukiah. Both vineyards are hillside, southwest facing slopes with excellent drainage. Both are approximately 25 years old and carry relatively small yields (2-3 tons per acre.)
Composition: 100% Zinfandel
Vinification and Maturation: After harvesting the grapes at peak ripeness, the fruit is hand sorted, destemmed and gently crushed, then left to cool soak on the skins for five days. Following the cool soak, some lots are inoculated with cultured yeast. Some lots are left to ferment with their native yeasts. The wines are pressed after ten to fifteen days skin contact, directly to barrel. Aging is in small, French and American oak barrels.

Related information:
Alcohol: 14.3%   **Brix**: 23.6   **pH**: 3.70   **TA**: 640 g/100ml
Bottling Date: December 1999
Production: 218

Winemaker's Notes:
The nose if filled with vanilla and black cherries with hints of smoke, pepper and spice. The mouthfeel is medium in weight with nice balance and a supple finish. This wine leans towards elegance instead of power and should pair wonderfully with roast chicken or grilled meats.

1998 Zinfandel, Dry Creek Valley

Appellation: Dry Creek Valley
Vineyards: The base wine for this blend is from a fifty-year-old vineyard in the heart of Dry Creek Valley that was planted in the forties. The dry farmed, head trained vines yield grapes that reflect the local terroir as well as any vineyard. The dusty aromas and raspberry and black cherry flavors are very typical for the area.
Composition: 94% Zinfandel, 6% Petite Sirah

Vinification and Maturation: Relatively the same for all Tarius Zinfandels.
Related information:
Alcohol: 14.1% **Brix**: 22.1   **pH**: 3.42   **TA**: 732g/100ml
Bottling Date: December 1999
Production: 338

W i n e m a k e r ' s   N o t e s :
The aromas are true to form for Dry Creek Valley. They include raspberries and black cherries with a dusty edge accented by spice and pepper. In the mouth the wine is medium bodied with a lush texture and a medium length finish. This wine should pair well with grilled meats and game birds like duck or pheasant.

1 9 9 8   Z i n f a n d e l ,   A l d i n e   V i n e y a r d

Appellation: Mendocino
Vineyards: The Aldine Vineyard is on the East side of Ukiah with a southwest-facing slope that has great drainage and good sun exposure. Yet, the cool Mendocino fall weather gives these grapes great hang time allowing extra time to develop flavors without getting sugars too high. This unique vineyard creates grapes with great balance and wonderful intensity of fruit.
Composition: 100% Zinfandel
Vinification and Maturation: Relatively the same for all Tarius Zinfandels. Aging is in small, French and American oak barrels.

Related information:
Alcohol: 15.3% **Brix**: 24.3   **pH**: 3.50   **TA**: 653g/100ml
Bottling Date: March 2000, released in the fall of 2000
Production: 247

W i n e m a k e r ' s   N o t e s :
Currently the nose is loaded with raspberries, wild blueberries and spices including cinnamon and vanilla. It is thick and rich in the mouth with silky tannins that coat the mouth, followed by a long, intense finish. This wine is almost too hedonistic to pair with food. I suggest drinking it by itself or with cheese!

The Lucas Winery

The Lucas Winery
18196 North Davis Road
Lodi, CA 95242
Phone 209-368-2006

Located in the Lodi Appellation, The Lucas Winery produces estate grown Zinfandel from 77 year old vines. A visit to this winery is not to be missed by any dyed in the purple, true Zinfandel fan.

Tasting and tours are private and require an appointment. These tours are given by David Lucas and are very extensive and educational. Open Saturday and Sunday, 12:00 to 5:00 for tasting.

Proprietor, Winemaker and Grape Grower: David Lucas

W i n e m a k i n g   P h i l o s o p h y: It is simple: It is hands on. From the vineyard to the barrel to the bottle, each step is taken by David Lucas. For those who are not seeking out the big names and are looking for a truly artistic, hands on approach to the experience of wine, then this winery and winemaker is for you. A visit and tour is not to be missed, not just by Zin fans, but by any wine lover. Tours start in the vineyard, looking at the practices that influence wine style, then off to the grand chai to learn about wine making practices and the importance of barrel selection. It is truly hands on!

1998 Lucas Estate Zinfandel

Appellation: Lodi
Vineyard Designation: ZinStar Vineyard 3.5 acre, 77 year-old vines.
Composition: 100% Zinfandel
Vinification and Maturation: Hand harvested two times. Open top fermentation, natural yeast, extended fermentation twenty five days, punch down. No filtering. Aged in French Oak for 12 months.

Related information:
Alcohol: 13.6%	**Residual sugar**: Dry
Brix at Harvest: 23.5	**Bottling Date**: 1/1/00
Release Date: October, 2000	
Production: 1000 cases sold at winery and distributed.	
Retail: $17.50	

Winemaker's Notes: The 1998 is a well evolved, balanced wine. Not over ripe, the fruit flavor is of fresh black berries, cranberry and spice with an elegant note of oak. The tannins are well rounded with a slightly aggressive mid-palate one expects from the ZinStar vineyard. A gentleman of pleasure. The 1997, which is also available, is an elegant, succulent, feminine Zinfandel with gentle structure and silky, almost secret tannins.

Food pairing: B.B.Q., fish or lamb. Excellent to deglase and serve with fresh raspberries and ice cream ...yum, you bet! Both wines are very enjoyable all on their own.

The Lucas winery also produces Chardonnay.

ZinStar Vineyard

LUCAS

1998

Lodi Appellation

ZINFANDEL

ALCOHOL 13.6% BY VOLUME

Lucas Vineyards, Lodi

Tobin James Cellars

Tobin James
8950 Union Road
Paso Robles, CA 93446
Phone: 805-239-2204 Fax: 805-239-4471

Tobin James Cellars, located at the corner of Highway 46 and Union Road in Paso Robles, is redolent with history from the old west. In the tasting room you are served wine on a bar that was built in the 1860's and is rumored to be from the favorite tavern of Jesse James. Also on site is a restored stage coach stop from the 1880's. The grapes are harvested from fifteen acres of estate vineyards plus purchased from other well known vineyards to produce some of the most exceptional wines anywhere. Tobin James has gained quite a reputation for our Zinfandels and visitors to the winery can find up to four Zinfandels available. Our other wines are equally incredible.

Tasting Room and picnic area open daily from 10:00 am to 6:00 pm. Mail order. Wine Club. Self guided tours.

Proprietors and Winemakers: Tobin James and Lance Silver

Winemaking Philosophy:

Due to the blending of many of the areas best Zinfandel vineyards, Tobin James produces Zinfandels that are consistent year to year. The Zins have tons of ripe, jammy fruit, full of blackberry and current with hints of spice and pepper. Tobin James is always true to his winemaking creed; "No matter how a wine turns out, that is how we planned it."

The Vintage:

1998 was a difficult vintage with some vineyards producing mediocre fruit and others that were wonderful. We sold the mediocre and blended the best!

1998 Zinfandel "Renegade"

Appellation: Paso Robles
Vineyards: No designation. 50% of the grapes come from Dusi Vineyard with vines that are 88 years old, head pruned and dry farmed. 25% from Bordonaro and the balance from a secret vineyard of 28 year old vines.
Composition: 100% Zinfandel

Vinification and Maturation: Hand picked and hand sorted grapes. Open top, slow fermentation seven to ten days. Gentle pump overs. Four rackings and filtering. Fourteen months in French and American oak, some new, some two to four years old. Six months of bottle aging.

Related information:
Alcohol: 14.6%   **Residual sugar**: Less than 0.2%
Brix at Harvest: 24.5   **Harvest Date**: 11/05/98
Bottling Date: 5/01/00   **Release Date**: 6/15/00
Production: 3000 cases sold at winery and distributed.
Retail: $16.00

W i n e m a k e r ' s   N o t e s :
The "Renegade" is just as its name implies: Big, Bold and a little Naughty. This is a ripe, juicy, fruit and spice packed wine with a lovely peppery finish. A versatile wine, it pairs with most anything but is at its absolute best on a starry night.

Tobin James also produces Chardonnay, Syrah, Cabernet Sauvignon, Cabernet Franc, Merlot, Barbera, Sangiovese, Late Harvest Zinfandel, Sauvignon Blanc and Petite Sirah..

Trentadue Winery

Trentadue Winery
19170 Geyserville Ave.
Geyserville, CA 95448
Phone: 707-433-3104 Fax: 707-433-5825

    The winery was founded in 1958 and bonded in 1969 by Leo and Evelyn Trentadue. Many old vines were on the ranch at that time and were left alone. Those vines are still in production today.

Tasting room at 320 Center Street, Healdsburg is open daily from 10 - 5.

Proprietors: Leo and Evelyn Trentadue   **Winemaker**: Miro Tcholakov

Winemaking Philosophy:
We try to pick when the fruit has the best flavors rather than a certain sugar level. We strive to work the wine as gently as possible and to do want the wine wants.

1998 Trentadue Zinfandel

    Appellation: Sonoma County
    Composition: 95% Zinfandel and 5% Petite Sirah.
    Vinification and Maturation: Hand picked, fermented in stainless steel, low pressure pump overs, with malolactic completed in barrels. Fermentation takes about six to seven days. Light filtration prior to bottling. Aged eleven months in one to three year old American and French oak barrels and six months bottle age prior to release.

    Related information:
  Alcohol: 13.85%   **Residual sugar**: 0.28%
pH and TA: 3.73/ 0.60grams/100ml
Harvest Date: October 15th, 1998
Bottling Date: September, 1999
Production: 590 cases sold at winery and distributed.

Winemaker's Notes:
Our 1998 Zinfandel shows a beautiful bright purple color with enticing aromas of red raspberries and spice. Rich and full in the mouth with deep flavors of black cherry, chocolate are framed by a delicate touch of vanilla. Lush and full-

bodied in texture with a long, spicy finish, this Zinfandel will improve with age but is delicious now with grilled sausages and full flavored dishes.

Trentadue also produces Carignane, Cabernet Sauvignon, Petite Sirah, Merlot, Sangiovese, Merlot Port and Petite Sirah Port.

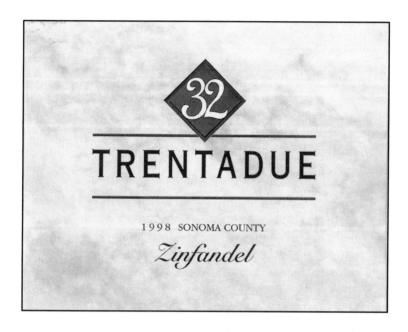

Tria

Tria
21885 8th Street East
Sonoma, CA 94581
Phone: 707-938-8890 Fax: 707-938-8889
Email: info@triawine.com
Website: www.triawine,com

Tria is a boutique winery within Napa Wine Company in Oakville, Napa Valley.
Tasting room open daily, by appointment only. Tours by appointment only
Owners: William Knuttel and Philip Zorn
Winemakers: William Knuttel, Philip Zorn and David Lattin
Business Manager: Debra Blodgett

Winemaking Philosophy:
Two principles guide all winemaking at Tria. First, the expression of true varietal character is preserved through the minimal application of traditional Old World winemaking techniques. Second, wine is to be enjoyed with food, and therefore must be balanced in all respects, including flavors, aromas, and textures as well as acidity, alcohol, and oak, guaranteeing rich, concentrated, age-worthy wines.

Vintage Notes:
The 1998 vintage was exceedingly cool, resulting in low yields and a late harvest. Hang times were extended well into October, and flavor maturity coincided with sugar maturity, a rarity for Zinfandel.

1998 Dry Creek Zinfandel

Appellation: Dry Creek
Vineyards: Piccetti Vineyard, augmented with Sbragia Vineyard and Bartolosi Vineyard. The Piccetti Family, who own the Columbus Salami Factory in San Francisco, grow most of Tria's Dry Creek Zinfandel on well drained hillsides of shallow, volcanic soil. In 1998, in the face of extremely low yields at the Piccetti Vineyard, additional fruit was sourced.

Composition: Zinfandel with minimal additions of other reds to insure a seamless middle body and finish.

Vinification and Maturation: Fermented with both native yeasts and the Rhône isolate D254 in closed-top fermenters with frequent, gentle pumpovers. Macerated 11-15 days. Malolactic in barrel. Sixteen months in predominately French oak, approximately 30% new. Racked every four months. No fining.

Related information:
Alcohol: 13.9%
Residual sugar: Dry
Harvest Date: October 7th, 1998
Bottling Date: March 6th, 2000
Release Date: Fall, 2000
Production: 1050 cases
Retail: $18.00

Winemaker's Notes:
Our 1998 Dry Creek Valley Zinfandel is a quintessential take on the appellation's bright berry and dried flower characteristics. The wine is deeply colored and, as a result of the cool vintage conditions, exceedingly rich with Zinfandel fruit. This is another in a long string of fantastic vintages in the Dry Creek Valley, and should age as well as all of the Tria Dry Creek Zinfandels from 1994 through 1997, which are maturing splendidly due to their inherent balance and concentration of fruit.

Tria also produces Pinot Noir and Syrah.

Truchard Vineyards

Truchard Vineyards
3234 Old Sonoma Road
Napa, CA 94559
Phone: 707-253-7153 Fax: 707-253-7234
Tours and Tasting by appointment only.

Truchard Vineyards is a small, family owned winery located in the Carneros region of the Napa Valley. Tony and Joann Truchard planted their first vineyards in 1974 and earned a reputation as excellent growers by selling their fruit to prominent Napa Valley wineries. In 1989 they decided to make wine for themselves, using their own grapes. The Truchard Estate Vineyard consists of 250 contiguous acres, and currently grows seven different grape varieties.
Owners: Tony and JoAnn Truchard
Winemaker: Sal Delanni

Winemaking Philosophy:
Truchard wines are made with the vineyard in mind. All of the wines are hand-crafted using traditional methods, then barrel-aged in underground storage caves. We strive to express the unique diversity and quality fruit of the Truchard Estate Vineyard. Our Zinfandel displays the exotic fruit and spice characters found in cooler-climate (Carneros) fruit.

1998 Truchard Zinfandel

Appellation: Carneros, Napa Valley
Vineyard: Zinfandel grows in the "Highlands" area of the Truchard Estate Vineyard. This area is dominated by austere hillsides composed of volcanic rock and ash.
Composition: 80% Zinfandel, 20% Cabernet Sauvignon
Vinification and Maturation: Gently crushed and destemmed into tanks. Pump-overs twice daily, with temperatures peaking at 88°F. 10 to 14 days skin contact. Aged for 10 months in both French and American oak barrels, of which 25% is new oak. Minimal racking and fining.

Related information:
Alcohol: 14.1%   **Residual sugar**: Dry

Brix at Harvest:24.0   **Harvest Date**: 11/02/98
Truchard Vineyards, Carneros Napa Valley   306

Bottling Date: August 1999   **Release Date**: February, 2000
Production: 488 cases distributed and sold at the winery
Retail: $24.00

W i n e m a k e r ' s   N o t e s :
Cooler temperatures throughout the growing season led to a late harvest. The extremely long hang time resulted in wines with complex aromas, intense flavors, and excellent acidity.
Rich aromas of black cherry, boysenberry, and cranberry, accented with cedar, tea and mint. The mouth is smooth and viscous - red berry/plum jam, vanilla, and ripe cherry, followed by a hint of cracked black pepper. Balanced acidity and subtle tannins life the palate and prolong the spicy finish.

Truchard Vineyards also produces Chardonnay, Pinot Noir, Syrah, Merlot and Cabernet Sauvignon. Production is 11,000 cases.

Truchard Vineyards, Carneros Napa Valley   307

Unti Vineyards

Unti Vineyards
4202 Dry Creek Road
Healdsburg, CA 95448
Phone: 707-433-5590 Fax: 707-433-9039
Email: mick@untivineyards.com
Website: www.untivineyards.com

Owned and operated by George and Linda Unti and their son, Mick Unti and daughter-in-law, Kimberly. Mick manages all winemaking, sales and marketing for the winery. Unti Vineyards consists of 64 acres planted to Zinfandel, Syrah, Sangiovese, Petite Sirah, Merlot, Grenache, Mourvèdre, Barbera and Dolcetto.

The 1998 Vintage:
A "Farmers Vintage" or tell me again, why we planted this grape?

Growing Zinfandel in 1998 was a bit like playing poker with nature. In mid-August it looked like we had been dealt a lousy hand. Many of the bunches were still green. Time for some difficult decisions, which to a Zin grower means crop thinning. Dad and his crew first went through the vineyard removing wings. (The classic Zinfandel grape bunch has a cute little wing of grapes at the top. This wing can inhibit skin development for the rest of the bunch and it's the ideal spot for rot to develop.) By early September we were still seeing some pink berries and the sugars were barely reaching 18° Brix, so, off with more crop. Unusually cool weather later in the month kept things from moving along. We were then faced with a mid-October harvest—tantamount to letting my two year old climb on the railing of our deck. Sure enough, a couple wet systems blew through, forcing us to drop crop two more times before harvesting our Zin at 23.5° Brix on October 21st. We'd be lying if we told you we weren't considering picking a week earlier in fear of losing the crop. But that's the fun of growing Zin.

I like to call this wine the Tim Hudson of our 1998 line up. Like the Oakland A's star rookie pitcher our Zinfandel arrived mid season in All Star form. The '98 always had more weight than our first vintage, but the fruit seemed to be buried under a sea of oak. In June we blended in 5 barrels of Syrah and whammo, the wine was lush, textured and tasted like blueberry jam. I heard a Tom Cruise-like voice coming from one of the Zin barrels saying "you complete me," to the Syrah. This is classic, fruity Dry Creek Zinfandel. The '98 has a little more oak than the '97, but because it's a bigger wine, it wears it well. Okay, I guess it's worth the trouble after all. But I don't know if my Dad is convinced. 1210 cases produced.

1998 Unti Zinfandel

Appellation: Dry Creek Valley
Vineyard: This Zinfandel comes from five acres of five-year old vines on our Dry Creek Bench ranch. It is an old clone Zinfandel producing small berries and short bunches. The vines are cordon-trained on a vertical trellis system to assist ripening and minimize bunch rot.
Composition: 81% Zinfandel 11% Syrah 8% Petite Sirah
Vinification and Maturation: Grapes were hand sorted, de-stemmed but not crushed and put into open and closed top stainless steel fermenters. After pre-fermentation soak, the must was inoculated with Bordeaux yeast, pumped over twice daily, and pressed immediately upon dryness to retain fruit. The wine was aged in 85% French (25% new) and 15% American (30% new) Oak barrels for 12 months. Syrah and Petite Sirah were blended into the Zinfandel for color and complexity. The wine was bottled unfiltered.

Related information:
Alcohol: 14.2%   **Residual sugar**: Dry
pH: 3.67   **Total Acidity**: 0.586 gm/100ml
Harvest Date: October 23rd, 1998
Bottling Date: 9/22/99
Release Date: 02/01/ 2000
Production: 1210
Retail: $18.00

W i n e m a k e r ' s   N o t e s :
Following this year's Zap tasting, the early word on the 1998 vintage is mixed. The opinion shared by a few reviewers is that the '98's lack the body of the preceding vintage. This comes as no surprise to me, since, as I mentioned, it was a difficult (and I'm being polite) growing season for Zinfandel here in the North Coast. I'm pretty sure our '98 Zin would have conformed to this generalization had we not dropped crop as early and often as we did. Though both the '97 and '98 are made in the same "fruit forward" Dry Creek style, in side by side

tastings, the '98 wins out every time, despite the fact that it is only now recovering from bottling and showing its stuff.

While farming practices like crop thinning can mitigate a tough Zinfandel vintage, blending other varietals can also help. As most of you know, we have a nice ace reliever in our bullpen called Syrah, which can really add depth and richness to Zinfandel. And remember, 1998 was just fine for our Syrah in Dry Creek. In fact, the Syrah was so fruity in '98 that we were able to blend 11% into the Zin without sacrificing our Dry Creek "fruit" Zin style. Now I know how Joe Torre must feel having Mariano Rivera around.

Unti Vineyards also produces Syrah and Sangiovese.

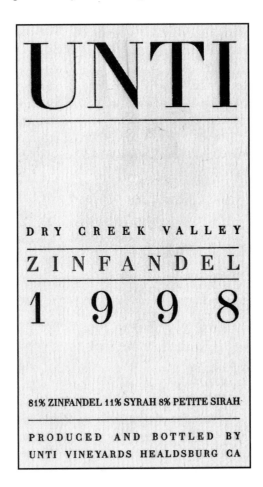

Unti Vineyards, Dry Creek Valley

White Oak Vineyards and Winery

White Oak Vineyards and Winery
7505 Hwy 128
Healdsburg, CA 95448
Phone: 707-433-8429 Fax: 707-433-8446

Established in 1981, White Oak Winery is dedicated to producing small quantities of distinguished lots. These wines reflect the uniqueness of the vineyards and climates of each appellation. Located in the Alexander Valley, the winery is surrounded by 17 acres of Merlot and 60 plus year old Zinfandel vines. Tasting hours 10 to 5 daily
Winemaker: Steve Ryan

Winemaking Philosophy:
The White Oak Zinfandel is picked at the optimum ripeness and hand sorted. Crushing of the berries is very minimal. Indigenous yeast slowly multiples creating longer and cooler fermentation. Steve gently pumps over twice a day for 21 days. Then pressed directly to barrel where it will complete malolactic fermentation. This White Oak Zinfandel is aged in a combination of American and French oak for 10 months and bottled unfined.

1998 White Oak Zinfandel

Appellation: Alexander Valley
Vineyard: Estate Alexander Valley vines planted in 1939 and 1925

Related information:
Alcohol: 14.5%   **Residual sugar**: Dry
Brix at Harvest:24.8   **Harvest Date**: 10/26/98
pH: 3.55   **TA**: .68   **Release Date**: July 2000
Production: 400 cases sold at the winery, through the wine club and limited distribution.
Retail: $18.95

Winemaker's Notes:
Our estate old vine Zinfandel characteristics start with a perfumed nose of red fruits, rich black berries, and soft tannins for a full bodied wine.

Wild Hog Vineyard

Wild Hog Vineyard
30904 King Ridge Road
Cazadero, CA 95421
Phone: 707-847-3687

Back in the hills between Fort Ross and Cazadero on the rugged Sonoma coast is this small family owned and operated winery and vineyard specializing in red wines. As certified organic grape growers, we take great pride in the quality of our grapes. This care and commitment is blended into each bottle of wine.

Tastings by appointment only. Mail order.

Proprietor: Daniel and Marion Schoenfeld
Winemaker: Daniel Schoenfeld

The Vineyards and the Winery:
Wild Hog Vineyard is at the headwaters of Wild Hog Creek 45 minutes west of the small town of Cazadero. At 1400 feet elevation and 5 miles from the Ocean east of Fort Ross, the vineyard avoids most of the summer fog. The climate of the coastal range, in conjunction with our farming practices, helps to produce flavorful grapes.

The Wild Hog farm is approximately 110 acres that encompasses a large timber house and a 2000 square foot winery - both powered by solar and hydro electricity, a three acre family garden and fruit orchard, and 5 acres of certified organic grape vines. Hot days, cool nights, and large rainfall produce excellent growing conditions and transform the rugged hillside into lush gardens and delicious fruits.

We are dedicated to farming and gardening without the use of chemical additives, insecticides, or herbicides and have been certified organic by California Certified Organic Farmers since 1981. A permanent cover crop of clovers and grasses is grown between the rows to hold and feed the soil and provide an environment for beneficial insects. We hoe and weedeat under our vines. Due to the excellent health of the vines, we have no significant insect problems. Our vines are fed compost and organic nitrogen as needed. A CCOF approved micronized wetable sulfur is used to control powdery mildew in the vineyard. Drip irrigation is used on young vines, but once established, they are dry farmed.

Wild Hog vineyard farms 1½ acres of Zinfandel grapes which produces approximately 4 tons per acre and 3½ acres of Pinot Noir which produces 2 tons per acre. The vines are vertically trellised with spacing varying from 8' by 12' to 7' by 9'. The vineyard is netted to keep out the birds and is fenced for deer and hog protection.

Wild Hog Vineyard is proud to be one of the few wineries from this unique wine growing region that is slowly but confidently becoming known for producing world class wines.

Winemaking Philosophy:
Believing the vineyard is most important, I keep an eye on everything while letting mother nature do her job.

1997 Estate Bottled Zinfandel

Appellation: Sonoma Coast
Vineyard Designation: All grapes are from Wild Hog's organic vineyard, five miles from the ocean.
Composition: 100% Zinfandel
Vinification and Maturation: Hand picked, destemmed, inoculated with a Bordeaux red yeast and fermented in small open top fermenters for an average of ten to twenty days. Gentle bladder pressing, racked three times on average, no filtration or fining. 19 months in mostly older French and some newer American Oak. We use about 1/5 new oak, preferring to let the fruit speak.

Related information:
Alcohol: 15.2%	**Residual sugar**: 0%
Brix at Harvest: 26.	**Harvest Date**: September 1997
Bottling Date: 6/02/99	**Release Date**: 6/30/99

Production: 200 cases sold at winery and distributed.

Winemaker's Notes:
Big and jammy with complex flavors of pepper, spice, black raspberry, and dark cherry. A monster wine at release, it has mellowed into a big, well balanced Zinfandel.

1997 Porter Bass Vineyard Zinfandel

Appellation: Russian River
Vineyard Designation: Porter Bass
Located in high above Guerneville between the cool climate appellation of Green Valley and Russian River Valley, Porter Bass Vineyard is owned by Dirck and Sue Bass. At an elevation of 500 feet, the hillside vineyard is sustainably farmed, head pruned, and produces approximately 2½ tons of Zinfandel per acre. The grapes, loaded with big fruit, strong tannins and good acid levels, consistently make a wine that ages well.
Composition: 100% Zinfandel
Vinification: Hand picked, destemmed, inoculated with a Bordeaux red yeast and fermented in small open top fermenters for an average of ten to twenty days. Gentle bladder pressing, racked three times on average, no filtration or fining.
Maturation: 19 months in oak barrels; of which approximately 4 months was in new American Oak.

Related information:
Alcohol: 15.3%   **Residual sugar**: 0%
Brix at Harvest: 26.   **Harvest Date**: Sept 30, Oct 1,1997
Production: 650 cases sold at winery and distributed.

Winemaker's Notes:
This wine has all the components: fruit, acid, ad tannins, to be a wonderful full bodied Zinfandel. If you like full bodied, big Zins packed with everything, this is it.

1998 Mohr-Fry Ranch Zinfandel

Appellation:
Vineyard Designation: Mohr-Fry
Composition: 100% Zinfandel
Vinification and Maturation: The grapes were put through the destemmer into several small fermentation tanks and Bordeaux yeast was added immediately. After 10 days, the must was pressed and the wine put into American oak barrels for 1 year.

Related information:
Alcohol: 14.7%   **Residual sugar**: 0%

Wild Hog Vineyard, Sonoma Coast   314

Brix at Harvest: 25.5 **Harvest Date**: October 2, 1998
Production: 300cases sold at winery and distributed.

Winemaker's Notes: Smooth well balanced, good fruit, easy tannins.

Wild Hog also produces Pinot Noir and Carignane.

PORTER-BASS VINEYARD
1998 ZINFANDEL
RUSSIAN RIVER VALLEY

PRODUCED AND BOTTLED BY WILD HOG VINEYARD
CAZADERO, SONOMA COUNTY, CALIFORNIA
ALCOHOL 14.4% BY VOLUME. CONTAINS SULFITES

GOVERNMENT WARNING: (1) ACCORDING TO THE SURGEON GENERAL, WOMEN SHOULD NOT DRINK ALCOHOLIC BEVERAGES DURING PREGNANCY BECAUSE OF THE RISK OF BIRTH DEFECTS. (2) CONSUMPTION OF ALCOHOLIC BEVERAGES IMPAIRS YOUR ABILITY TO DRIVE A CAR OR OPERATE MACHINERY, AND MAY CAUSE HEALTH PROBLEMS.

ZAP
Zinfandel Advocates and Producers

ZAP
P.O. Box 1487
Rough and Ready, CA 95975
Phone: 530-274-4900 Fax: 530-274-4904
Email: zaprr@oro.net
www.Zinfandel.org

In 1991, with just a small band of 25 wineries, ZAP set out on a mission. The goal was, "To bring together those dedicated to Zinfandel, knowledgeable about its virtues and willing to spread the word." Today, ZAP membership includes 230 wineries, 6 associate members and over 5,500 Zin-fiends and the numbers are continuing to grow. There is no other variety that commands this kind of passion or loyalty. But then again, there is no other variety like Zinfandel. It is truly America's treasure.

Quarterly Newsletter, special events, member discounts.

Executive Director: **Rebecca Robinson**

Philosophy:
The Association of Zinfandel Advocates and Produces (ZAP) is a non-profit organization. ZAP is dedicated to educating the public about the Zinfandel grape. It has a unique place in American culture and winemaking. Zinfandel's distinctive character and exceptional quality place it among the prominent wines of the world.

ZAP Zinfandel Advocates and Producers

Appellation: California
Composition: Over 230 Zinfandel producers, 6 associate members and 5,500 advocates.
Vinification and Maturation: ZAP will continue to grow and evolve, promoting Zinfandel and its place as a national treasure.

ZAP Notes: For Zinfandel lovers, there is no better organization to be a part of. Informative newsletters, special events at wineries, not to mention the annual gathering of Zinfandel producers at the Festival Pavilion, Fort Mason in San Francisco every January. This is the largest tasting of its kind and one not to be missed. Zinfandel tastings are expanded beyond the borders of California for those who do not have the pleasure of living in Zin-land.

ZAP also promotes the Heritage Vineyard: a project that includes the identification, planting, promotion and scholarly pursuits related to historically important Zinfandel grapevines. Adopt-A Vine Program was created to give ZAP members a chance to get directly involved in the Zinfandel Heritage Vineyard Project.

Glossary

••A••

ACID AND ACIDITY

Some acids, such as malic and tartaric, are naturally present in grapes while others, such as lactic acid are formed during fermentation. Acids contribute to the taste of wine giving it a fresh, crisp, tart flavor and heightening the fruitiness. The level of acid is important for balance, to help preserve color and aid in the aging ability of wine. Excessive amounts of acids can cause the wine to taste sour or feel sharp, leaving a harsh sensation in the mouth.

BLENDING

Blending is a common practice and can incorporate many different techniques. For example, a wine can be a blend of various vineyard sites with each vineyard relating its own characteristic to the aromas and flavors of the wine. Some winemakers may age a certain percentage of the wine in new oak and another portion in older oak that is relatively neutral in an effort to control the influences the oak imparts to the wine, later blending those wines before bottling. There are many types of blending, but we generally think of blending as the combination of different grape varieties. Blending grape varieties, vineyard sites, separate vinification processes, or finished wine from different barrels or steel tanks, are common winemaking techniques.

BODY

Body is the sensation of fullness or weight of the wine in your mouth. Alcohol, acid, extract (dissolved solids) and glycerol are some of the components that give a wine body.

BOTTLE AGING

Wine continues to mature through the process of bottle aging. Harsh tannins soften, the bouquet evolves and a richness and complexity develop as the nuances of flavors and aromas interact. Some wines do not require extended periods of bottle aging as they are meant to be consumed in youth.

BOUQUET

Not unlike a mixed bouquet of flowers, the bouquet of a wine incorporates many fragrances. Each component such as varietal aromas and the aromas that develop from oak aging, malolactic fermentation, etc. are "married" through the process of aging. This combination of fragrances is the wines bouquet.

BRIAR
The tactile sensation of a wine that is prickly like the black pepper-like flecks felt on the tongue, a common Zinfandel trait.

BRIX
A measurement of the concentrated sugar in grapes, must or wine. This is one of the methods winegrowers and winemakers use to determine the desired ripeness of the grapes for harvesting.

BUTTERY
A tasting term describing the aroma or texture of creamy butter. A wine that has undergone malolactic fermentation and, or oak aging will impart these nuances.

••C••

CANOPY MANAGEMENT
Shoots and leaves form the canopy of a vine and it is these leaves that protect the grapes from over exposure to sun and wind. However, if there is not enough exposure the grapes do not ripen, develop disease or produce low yields. There are several techniques utilized to alter the canopy and the fruits exposure to the elements including trellising, leaf pulling, pruning, shoot thinning and trimming. Canopy management is critical to the development of quality fruit and will vary depending on the variety and vineyard. The complexity of canopy management can not be covered in this text but is worth further study for those interested in viticulture.

CAP
During red wine fermentation grape solids consisting primarily of skins along with seeds, and stems, form a tight mass that floats on the surface of the juice. This mass is called the cap and to extract color, flavor and tannins from the grape solids, this cap must be bro-

ken up and submerged into the juice. **Cap management is a selective process. Punching down or pumping over are two of the most common methods used. Sprinkler systems are used by some while still others utilize a screen to keep the cap continually submerged. There is also a mechanical method of rotating the juice and cap. The method chosen is important as is the frequency of these methods which are performed during fermentation to achieve a desired level of extraction. Keeping the cap moist not only facilitates extraction but keeps bacteria from forming on the surface of the cap.

CARBONIC MACERATION
A process in which whole berry clusters are piled on top of one another with the weight of the top layers acting as a natural press, crushing the grapes below and releasing their juice. Fermentation begins, releasing carbon dioxide which envelops the upper layers of grapes blocking out oxygen. The upper layer of grapes remain unbroken and begin fermentation within the solid grape. The middle layer, still unbroken and surrounded by juice, also begins fermentation but at a slower rate. This gentle maceration process produces an aromatic, light bodied, fruity, low tannic, brightly colored wine. Winemakers may choose to utilize this method with small lots of Zinfandel to add to the final blend.

CASSIS
The aroma and flavor of black currants.

CEDAR
Typically the result of oak aging this aroma is reminiscent of a cigar box or cedar chest. Cedar is also a characteristic of Cabernet Sauvignon or Petite Sirah and when blended with a varietal Zinfandel can bring this nuance to the wines bouquet.

CELLAR
A cellar is a temperature controlled environment where wine is processed, aged and stored. In a winery these areas are referred to as the barrel room or cellar. Once the wine is sold, maintaining the wine in a proper environment is very important because wine is a living

thing continuing to change and evolve in the bottle. Wine should be stored in an area protected from vibration, light, and fluctuations of temperature. The temperature can range from 45° to 65°F but should be constant. The bottle should also be stored on its side, or cork down to keep the cork from drying out. Home cellaring can be expensive but with time spent researching the proper conditions for cellaring, along with a creative mind, a closet or cupboard can be transformed into the perfect place to lay wine down.

CHARACTER
A distinctive quality that is recognizable, separating it from other features. The flavor, color, aromas and body of a wine can have characteristics that are discernible features of the variety, the vinification process, the appellation, the vineyard and even the winemaker.

CHAI
Pronounced shay, it is a French term meaning cave but in the U.S. can refer to the cellar where barrels of wine are kept under controlled temperatures.

CHERRY
The flavor and aroma of cherries ranging from tart cherry to ripe, rich black cherry is a very common characteristic of Zinfandel enhanced by specific growing regions and clones.

CHEWY
A wine that is full-bodied and tannic can sometimes be described as chewy. The wine is so mouth filling, the texture thick and rich that it seems the wine could be chewed.

CHOCOLATE
The flavor and aroma of rich, deep, bitter-sweet chocolate that can be a characteristic of the variety, the vineyard or a result of oak aging.

CLARET
The English used this term when referring to the wines of Bordeaux. and the blending of the five noble grapes; Cabernet Sauvignon,

Merlot, Cabernet Franc, Petite Verdot and Malbec. Thus, a wine that is blended can be referred to as a claret style. In Spanish, the word Clarete refers to a light hued, light bodied wine.

CLARIFICATION
A process that removes solids from the grape juice or new wine. One reason clarification is performed is for aesthetics purposes, removing any sediment so the wine does not appear cloudy or hazy. Clarifying also removes the new wine off of sediment that can impart harshness to the wine. Racking, filtering and fining are some of the methods of clarifying and are performed at various stages of the winemaking process. Barrel aging and stabilization will also facilitate clarification.

CLASSIC
Winemakers may use the term to denote a style of wine that remains fairly consistent in character from vintage to vintage. There are no regulations on the use of this term on a label.

CLONES
Propagation for certain perennial plants, like grape vines, can occur through cutting or grafting to insure genetic identity. A specific vine may be selected for its yield, lack of viruses, resistance to disease, compatibility with site, varietal characteristic, etc. and from this "mother vine" buds are taken. Clonal selection is an important study and worth further inquiry for those interested in viticulture.

COLD SOAKING
A maceration process performed prior to fermentation. The newly harvested, destemmed fruit is kept in temperatures precluding fermentation. This process can enhance the fruit character of the grape and heighten color extraction.

COLOR
The color of wine comes from the skin of the grape and is imparted to the wine through the process of maceration; keeping the juice or new wine in contact with the skin of the grape. Winemakers will control the contact with the skins to reach a desired hue. Blending and

oak aging can enhance and stabilize the color of a wine. Acid also plays a part in preserving color. Color is an indicator of the variety, style, quality and age of a wine. Young reds have a robe (color) of purple and crimson, as the wine ages or because of oxidation wines will turn more brick red. Zinfandel has a notoriously fugitive hue, losing its bright plumy color as it ages.

COMPLEXITY
Complexity is the intricate layering of flavors, sensuous textures and the subtle nuances of bouquet integrated in balance and harmony. Complexity defines a wine of interest and of quality.

COOPERING
Simply, it is barrel making, but there is nothing simple about the process, the craft or art of barrel making. How the barrel is made is just as important as the wood used. Coopering has a wonderful history dating back to Roman times and is a delightful subject to read about. If ever the opportunity to visit a cooperage arises, by all means do it. It is fascinating.

CROP THINNING
A common viticultural practice to ensure quality fruit. Farmers remove flower clusters or newly set clusters to reduce the amount of fruit thereby strengthening the vine and affording the fruit clusters that remain a better opportunity to develop and mature.

CRUSHING
One of the most widely used techniques, crushing is a mechanical method of extracting the juice from the grape berry. Equipment used today performs a continuous operation of gently breaking the skin and releasing the juice. Stems and leaves are separated out and the resulting liquid, along with the seeds and skins, are placed into fermenting tanks. In white wines, the seeds and skins are discarded.

CUVÉE
Pronounced koo-vay, this term when applied to still wine simply means a blend of different varieties or the grapes from various vineyards.

DELICATE

A wine that is light in style and body, not overly assertive, soft and pleasing is described as being delicate.

DEPTH

When describing wine the term depth is synonymous with complexity. Depth is also used to describe the intensity and richness of the wines color.

DESTEMMED

Once the grapes are brought in from the vineyards they are placed in a machine that gently separates the grapes from the stems and leaves. Crushing can also occur simultaneously. Many winemakers choose to only destem the grape clusters and leave them intact, uncrushed. See crushing and whole berry fermentation.

DRY

The opposite of sweet, wine that is dry is considered to have no more than 0.2% residual sugar. Dry is also a term that refers to the fermentation process. Winemakers will state that their wine is fermented dry, meaning fermentation is complete once all residual sugar is converted to alcohol.

DRY FARMED

This wine growing term refers to the lack of irrigation in the vineyard. Vines require a minimal amount of water for good fruit production. If the vineyard is not irrigated it means that a water source below ground is available for the vines. Zinfandel is often dry farmed as irrigation or too much water tends to cause over-cropping in this variety.

DEVELOPED

Wine that has developed has reached a level of maturity making it ready to be consumed. An under-developed wine is one that requires more aging or was not balanced to begin with.

EARTHY

The flavor or aroma of moist, rich soil, a damp forest or mushrooms. Earthy is typically a positive term with no reference to dirt.

ELEGANT

A style of wine that reflects a refinement, showing finesse and grace.

ESTATE BOTTLED

One hundred percent of the grapes in the wine must have come from vineyards that are within the same appellation and are either owned or held in long term lease by the winery that is producing the wine.

EXTENDED MACERATION

The extended period of time the wine remains in contact with the grape solids after primary fermentation has completed. Winemakers who use this process believe it softens the tannins, extracts the maximum color and flavors and creates a more complex wine.

••F••

FERMENTATION

The process of converting grape sugar into ethyl alcohol and carbon dioxide through the introduction of yeast: Grape Juice into Wine. Although the fundamentals of fermentation are not overly complex as fermentation will occur naturally but there are many variables that can effect the final product such as the rate of fermentation, temperature and selection of yeast. How fermentation is handled and in what vessel is a discerning process that depends upon the grape variety and a winemakers intent.

FERMENTATION TANK

Concrete vats, stainless steel tanks and wood barrels are some of the vessels utilized in the fermentation process. For home wine makers a garbage can is usually the most cost efficient.

FIELD BLEND

Once a common practice, winegrowers planted various grape varieties together within the same vineyard. The grapes were often harvested, crushed and fermented together creating a natural blend in the wine. Some of the older Zinfandel vineyards will have predominately Zinfandel with Alicante Bouschet, Carignane, Mourvèdre or a host of other varieties peppered here and there in the vineyard. Grape growers today tend to plant one variety within a block or vineyard site, leaving the blending up to the winemaker.

FILTERING

Typically performed prior to bottling, this clarification process screens out potentially harmful microorganisms, dead yeast cells, and solids from the wine. Some winemakers prefer to limit the degree of filtering, or not filter at all, believing that the process can diminish certain characteristics and that wine gains in complexity if bottled with a certain amount of sediment.

FINING

Particles that cloud the wine or can cause bitterness are removed through the clarifying process of fining. Fining agents such as egg whites act like a glue or a net capturing microscopic proteins. This sediment sinks to the bottom of the vessel where it is removed through racking or filtration. Some winemakers choose not to fine their wine for basically the same reason they select not to filter their wine.

FINISH

The combination of aroma, taste and texture of the wine that lingers in the mouth after the wine is swallowed. A silky finish, or a long lingering finish are some of the positive impressions that will be imparted and is a characteristic of a well balanced wine. A finish that is sharp, harsh, hot or nonexistence shows a wine that lacks balance and appeal.

FREE-RUN
A technique used to gently separate wine from fermented must after full extraction is completed in order to achieve a softer tannin content.

FRENCH OAK BARRELS
The most popular wood for barrel making comes from various appellations in France, although there are many sources for oak throughout Europe. Each region produces oak with different characteristics in tightness of grain, tannin, and flavor components, important attributes that will influence the wine. In general, French oak tends to be more subtle than American oak with rounder, softer flavors. Some common flavor and aroma characteristics of French oak barrels are vanilla, caramel, clove, nutmeg, green wood, resinous, hazelnut, roasted coffee, hickory smoke, and bacon.

FRUITY
A tasting term describing the fruit characteristic of a grape variety, typically found in young wines. Raspberry, cherry, dried fruit, blackberry are but a few of the more precise descriptors used.

••G••

GRAVITY FLOW
Some wineries will use the natural process of gravity when racking or bottling the wine. This practice incorporates the gentle pull of the earth's natural force eliminating any harsh or rough handling of the wine as well as reducing the amount of oxygen infused into the wine.

GREEN
Not a flattering term, it refers to a wine that is not just young, but immature. The wine tastes and smells unripe, undeveloped, and uninteresting. Green should not be confused with herbal or vegetal.

••H••

HAND PICKED AND HAND SORTED
A labor intensive practice performed by grape growers, vineyard crews, winemakers and/or cellar workers to ensure that only the best grapes go into the wine. Zinfandel can be prone to mildew, rot and

uneven ripening. Winemakers want those grapes discarded. Hand sorting extends the process, taking a second look at what is going into the destemmer or fermentation tank.

HANG TIME
Grapes that ripen slowly have more concentrated fruit character, good acid and tannin structure. A long growing season, or hang time, affords the fruit a slow, even ripening rather than a rush to maturity.

HARMONY
A lyrical term that translates well to wine and indicates a wine that is balanced.

HEAD TRAINING
Zinfandel is often adapted to this classic approach of vine training. The trunk takes on the appearance of a small tree with a defined head of old wood from which arms extend. From these arms new shoots will form from the budded spurs left by winter pruning. The choice of training influences vigor, yield, pruning, canopy management, and harvesting. Training also impacts the grapes susceptibility to mold, mildew, exposure to the elements, ripening and the list goes on. Some winegrowers do use trellising for Zinfandel and as with many wine-growing and winemaking techniques, it, too, has its advocates and its distracters. A topic worth further reading for those interested in viti-culture.

•• I ••

INKY
A color describing the deep, dense hue of a wine. Petite Sirah is often described as having an inky color.

INNERSTAVES
Barrels have a certain life span, not just in their ability to hold liquid, but in the flavors, aromas and tannins the wood imparts. Innerstaves are staves of new oak that are connected and look like a small, round fence. They are inserted into a barrel that has become neutral from use, although neutral barrels are often desired. Innerstaves are but

one method of extending the life of a barrel while managing oak flavors. Adding oak chips is another method of adding oak aromas but is considered an economic way of introducing oak to lesser quality wine.

••J••

JAMMY
Like a spoonful of Grandma's homemade jam, wines that are jammy have concentrated, full berry and fruit flavors.

••L••

LEES
After fermentation, dead yeast cells and small grape particles settle to the bottom of the tank or barrel. Left in contact with this sediment the wine gains in flavor and mouthfeel. Lee contact also encourages malolactic fermentation adding softness and complexity. The French term, Sur Lie, (pronounced soor lee) will often be stated on the label to indicate the winemaker has used this technique.

••M••

MACERATION
The process of keeping the grape juice in contact with the skins, seeds and stem fragments is called maceration. The purpose is to extract color and flavor compounds from these solids. There are three levels of maceration: cold soak which is a pre-fermentation maceration; pumping over or punching down are two examples of maceration that occur during fermentation; and extended maceration, the process that occurs after fermentation is complete. The period of maceration is determined by the variety and also the style of wine intended.

MACROCLIMATE (MACRO: LARGE, LONG OR GREAT)
A term that refers to the climate of a growing region such as California, or Sonoma County.

MALOLACTIC FERMENTATION
Also referred to as secondary fermentation, malolactic fermentation is a process that converts malic acid into lactic acid. Malic acid is natu-

rally present in the grape and is the same tart acid found in green apples. Lactic acid, a softer acid, is the same acid in milk giving it that creamy texture. In the simplest of terms, this tart green apple-like acid, malic, is converted to the soft, creamy milk acid, lactic. A by product of this process is the production of a chemical called diacetyl which gives wine that buttery smell. The result is a wine that is softer, and rounder with a creamy texture. Malolactic fermentation can occur naturally but some winemakers prefer to control it by inhibiting the bacteria that causes this fermentation and later inoculating the wine to induce the process.

MATURE
A mature wine has reached the height of its potential and is ready to drink. Some wines are mature as soon as they are bottled, others require bottle aging to attain maturity.

MEATY
Synonymous with chewy.

MESOCLIMATE (MESO: MIDDLE)
A term that refers to the climate of specific place or space. Mesoclimate is the climate of the vineyard site, such as a hillside vineyard or a vineyard situated on benchland.

MICROCLIMATE (MICRO: SMALL)
Microclimate is the climate within the vine and the soil. Changes in temperatures between rows of vines, even between grapes clusters is the microclimate. Unlike mesoclimate or macroclimate, microclimate can be altered and influenced by the techniques of vineyard practices such as row spacing or canopy management.

MUST
The unfermented juice of grapes as well as the seeds, skin and pulp is referred to as the must.

NEUTRAL BARRELS

A barrel becomes neutral typically in three to five years of use. At this stage the barrel still functions not just as a vessel, but retains much of its ability to "breathe" with the wine, the process of evaporation and slow oxidation that concentrates the wine. With a neutral barrel the flavors and aromas that are imparted by oak have all been leached out by the wine through time and use. This is not a negative attribute. Winemakers may select neutral barrels as part of the balance of the oak character they wish to impart to the wine. Blending wines that have been held in older or neutral oak with wines from barrels that are new, help mange the oaks impact upon the wine. Overly oaked wine is not a balanced wine and can even be a sign of a flawed wine that is hiding behind a wooden mask.

NOSE

Refers to the smell, aroma or bouquet of a wine.

OAK AGING

Utilizing oak barrels to age wine influences many aspects of the ultimate style of the wine. Because of the porous nature of wood micro-oxidation of the wine occurs, concentrating the flavors of the wine as it matures. The woods flavors and aromas are imparted and incorporated into the bouquet. The degree of toasting the barrel receives also convey varying characteristics to the flavor and aromas of the wine. The natural tannin of the wood interacts with the tannin of the wine allowing the wine to soften and gain complexity. All components round out and integrate with one another, becoming harmonious. Oak aging also helps to stabilize color in red wine. The shape of the barrel naturally clarifies the wine with all sediment settling to the bottom. Nuances of vanilla, smoke, butterscotch and coconut are just some of the characteristics of oak aging.

OAK BARRELS

The use of wood barrels is an ancient method of storage and transport but winemakers today have more to consider than just a place to store wine. Where the oak comes from, how it was seasoned (dried), who does the coopering, French oak, American oak, how much toast, new oak, old oak... the query goes on. Oak is the wood of choice because it is strong, yet pliable, and is a watertight wood with a natural composition that partners well with wine. The average life span of a barrel for wine production is about five years although some winemakers continue to use their barrels until they are structurally corrupted or have an undesirable bacteria that can not be eliminated. Winemakers may choose to use a barrel only once or use it longer than five years depending upon the variety or style of wine intended and the impact they want from the wood. Barrel sizes and styles vary but the most common is a 60 gallon barrel.

OAKY

Wine that has been aged in oak barrels will have toasty, vanilla flavors and aromas referred to as oaky.

••P••

PEPPERY

A wine tasting term for wines that have spicy, black or white pepper flavors and aromas.

PERFUME

Typically a term that refers to the fragrant smell of the wine but can also be used to describe a wine that has floral aromas.

PHYLLOXERA

This tiny aphid feeds on the roots of vines, eventually killing them. Once a vine has this nasty pest, there is nothing that can be done. For now, grafting the vine over to rootstock that is resistant to the louse seems to be the key in prevention. Replanting a vineyard is not only costly but vines do not produce wine quality fruit for three years or more. For a grape grower, phylloxera means the slaughter of his vines. They were first reported in England in the late 1800's but it was America that unwittingly exported them. In the latter part of the 19th century it almost devastated the vineyards in France, and it remains a threat in many grape growing regions. Some of the older plantings of Zinfandel have survived this pest because they are in relatively isolated areas, developed resistance or planted in sandy soil which this louse can not move through.

PIERCE'S DISEASE
Not a new disease, it is blamed for the destruction of vineyards in Alabama in 1821, in southern California in 1883-1886 where it was referred to as Anaheim Disease, and in the San Joaquin Valley in 1937-1944. Pierce's disease, named after Dr. Pierce who studied the disease in the 1880's, is a bacterial infection that grows in the water-conducting system of the vine. Symptoms are shriveled berries, scorched and deformed leaves and poor wood maturity with new growth stunted. At this point in time there is no cure and the vine dies. Although this disease is indigenous to the United States, affecting other crops such as almonds and alfalfa, what is new is the vector. Sharpshooters, or leaf hoppers are the carriers of this disease. The blue-green sharpshooter was the known culprit but its feeding pattern, population size and range of habitat allowed for some management. The new vector, the glassy winged sharpshooter is larger, a stronger flyer, can exist in a wider range of habitats, increased population size, and its feeding patterns make this insect a greater threat to the vineyards.

PLUM
A wine tasting term that refers to color as well as aroma and flavors.

POMACE
The seeds and skin of the grape after pressing. This pomace is often returned to the vineyard as compost.

PRESSING
A mechanical method of extracting the juice from the grape. In red wines it is typically performed after fermentation but winemakers

may select to press some of the wine during fermentation. Basket presses are the most traditional but uncommon with the tank or bladder press being more economical and considered to be a more gentle process.

PUMPING OVER

A mechanical process of maceration where the fermenting juice is drawn up from the bottom of the tank and pumped over the cap to extract more flavors, color and tannin. Because the cap floats on the surface of the juice and is partially exposed to air pumping over insures the cap does not dry out and develop unwanted bacteria growth. Used gently, some winemakers believe this process of pumping over typically results in a wine that is more approachable in its youth with a softer, more elegant tannin structure. There are those who disagree, believing pumping over is highly agitating to the young wine, and to a lesser degree, simply irrigating the cap, extracting very little. The debate goes on.

PUNCHING DOWN

This labor intensive form of extraction is performed by literally breaking up and pushing the cap down into the fermenting juice. A wooden paddle with a circular disc attached at one end is typically used although, depending on the size of the fermentation vessel, a two-by-four or your hands will do the job as well. Punching down has the same goal as pumping over: to extract flavor, color and tannin and to keep the cap moist. Some winemakers believe it is not as sensitive of an operation as pumping over, as punching down may increase the break up of the skin and seeds extracting more tannin which can result in wines that are belligerent and coarse. Still, others believe this method to be far less aggressive and harsh than pumping over, drawing more flavors and color from the skins and avoiding excessive aeration. It is an old world method, not text book, and for the winemakers who use this process, it is the only way.

RACKING

A process of clarification, racking removes the clear wine from the sediment or lees that have settled to the bottom of the barrel or tank. The wine is pumped or siphoned off these particles and placed in another container. This process is very labor intensive as each container must be thoroughly cleaned before any wine is returned to them. Racking not only helps to clarify the wine but because aeration (exposure to oxygen) occurs through the pumping process tannins become softened.

RASPBERRY

One of the more notable characteristics of Zinfandel, raspberry flavors and aromas are varietal as well as vineyard characteristic.

RESERVE

A wine labeled reserve does not imply the wine is of a better quality as there are no regulations that must be adhered to to use this term on a label. In some instances it can mean that specific grapes or vineyard sites were used because of their quality or that more labor and expense were incurred to produce this wine. If in doubt when purchasing a wine labeled reserve rely on the reputation of the winery and the winemaker.

RESIDUAL SUGAR

The percent of unfermented grape sugar in finished wine. When stated on a label it can help identify whether a wine will be dry or sweet. For example, less than 0.2% will indicate a dry wine, as the percent increases so does the level of sweetness perceived by taste.

RIPE

A tasting term that describes the fruitiness of the wine. Zinfandels are often characterized as having rich, ripe black cherry flavors. These intense fruit flavors indicate the grapes were harvested at their optimum ripeness.

ROUND

The sensation of a full mouth feel that a well balanced and full bodied wine will impart is referred to as round.

••S••

SECOND LABEL

Some wineries may choose to produce wines using a second label rather than their primary label. This is done for a variety of reasons and does not indicate a second rate wine any more than "reserve" guarantees a wine of heightened quality.

SEDIMENT

Solids, such as dead yeast cells that result from fermentation, or seeds, grape pulp and skin that do not dissolve are part of the fragments that make up sediment. Sediment can be removed from the wine through racking, filtering and fining. Wine that is unfiltered or unfined retain this sediment and when serving these wines it is best to decant to avoid a mouthful of grit. Sediment is harmless and winemakers who select non-filtering do so because they believe it enhances the wine.

SILKY

A wine tasting term referring to the smooth, velvety texture of a wine.

SMOKY

Smoky wood aromas are typically the result of barrel aging.

SOFT

A wine tasting term that refers to the fruit being pleasing and luscious. It can also describe tannins that have mellowed through oak or bottle aging and feel less astringent.

SPICY

A varietal character that includes pepper, cracked black pepper, nutmeg, clove, cinnamon, allspice, licorice and anise. Zinfandel tends to have a more pronounced pepper spiciness than other varieties, but appellation, specific vineyard sites, blending other varieties and contact with oak can impart and enhance varying flavors and aroma of spice in the wine.

STABILIZATION
A method of clarification utilizing temperature to remove unstable proteins that can later effect the wine and tartrates that can cause the wine to become cloudy or form harmless crystals.

STRUCTURE
Wines are made up of many components; acid, tannin, bouquet, flavors, oak influence, etc.. They are like building blocks, part of the composition and form that result in a well balanced wine.

SULFUR DIOXIDE
A common chemical compound used in the winemaking process to inhibit undesirable bacteria and wild yeast growth. SO2 also inhibits oxygen and encourages fermentation.

••T••

TANNIN
A natural preservative, tannic acid is what enables red wine to age. Naturally found in the seeds, stems and skins of the grape, tannin is extracted through maceration, alcohol fermentation and contact with oak barrels as wood also contains tannin. The dry, chalky, dusty sensation in the mouth is tannin and as the wine ages, this acid diminishes.

TERROIR
A French term that is difficult to translate because it encompasses so many factors of the vineyard. Terroir is the soil, the climate, drainage, exposure to sun, wind, fog, vineyard practices, the hand of the farmer, everything that influences the development of the vines, the grape and ultimately the wine.

TEXTURE
Texture or mouthfeel is the tactile sensation of the wine on the tongue or back of the throat.

TOAST

The inside of an oak barrel is toasted by fire during the process of barrel making or coopering. There are several degrees of toasting, ranging from light to heavy with each influencing the flavors and aromas imparted to the wine. Toasting provides a buffer between the woods tannin and the wine and softens the phenol extraction of the new barrel. The toasted barrel gives very distinct aromas reminiscent of fresh baked bread, butterscotch and toasted nuts.

TOASTY

An aroma or flavor imparted by oak aging or fermentation.

TOPPING UP

A winemaking regiment performed to replace wine that has evaporated in the barrel due to micro-oxidation. Bacteria love oxygen and if air space is left between the wine and the bung, unwanted bacteria develop and can potentially destroy the wine.

TRELLISING

A vine training method where cordons, arm-like branches that extend from the trunk, are trained along wines influencing vigor, exposure, yield and harvesting. There are many different styles of trellising and the choice will depend upon site, the cultivar or fruiting variety and farming practices. Some trellising systems are set up with movable wires to insure good sun exposure for the grapes and leaves.

••U••

UNFILTERED

Winemakers will purposely choose not to filter their wines feeling it adversely effects the color, flavor and body of the wine. It is often stated on the bottle that the wines have remained unfiltered therefore one can anticipate a bit of sediment in the bottle or the pretty rose colored tartrate crystals that attach themselves to the end of the cork.

UNFINED

If given enough time, proteins and sediment and even color will naturally drop out of the wine so some winemakers select not to fine their wine. It can be stated on the bottle that the wine is unfined and like unfiltered wine a bit of harmless sediment can be expected.

••V••

VANILLA
A wine tasting term indicating the flavor or aroma of vanilla beans imparted by oak aging.

VARIETAL
Often interchanged with variety, varietal is a labeling term indicating that a minimum of 75% (51% before 1983) of the wine in the bottle is comprised of a single variety. In other parts of the world the percentage or labeling practice may vary but any wine produced in the United States that labels their wine as a varietal must contain a minimum 75% of the stated variety. Winemakers are not required to state the variety or varieties that make up the other percentage of the wine.

VARIETY
Common grape varieties are Zinfandel, Chardonnay, Cabernet Sauvignon, Pinot Noir, etc. with the word variety referring to a single type of grape.

VARIETAL CHARACTER
Varietal character refers to the color, body, flavor or aroma that is unique to a specific grape variety. Each grape variety has characteristics that distinguish them from other varieties. Certain varieties have more pronounced and discernible characteristics than others and winemaking techniques as well as vineyard site and clones can influence these attributes.

VERASION
A term that defines the period of time when the grapes begin to ripen, showing their true colors.

VINE AGE
Vines that are safe from disease, natural or man-made destruction, continue to grow and produce fruit. As the vine ages, it loses vigor

and production lessens. What fruit the vine does produce tends to be more concentrated, with richer, deeper aromas and flavors. Zinfandel vines have achieved old age for a variety of reasons. One is that they were often planted on hillsides where other crops could not grow. When Prohibition came along, devastating the wine industry, these vines remained in place as it was useless to replant the vineyard to anything else and Prohibition did not prohibit winemaking for personal use. Many Italian immigrants continued to farm small blocks of Zinfandel for their own consumption as wine is, traditionally, a part of life. How grateful we should all be for such traditions.

VINEYARD DESIGNATION
To label the wine with a vineyard designate, 95% of the grapes that went into that bottle of wine must have come from the vineyard specified. Winemakers are selecting specific vineyards because of the uniqueness and quality of the grapes that those sites produce. A very focused "personality" emanates in a wine from a single vineyard. Even though the site may be one, the microclimate of that vineyard can be very diverse. Some winemakers, on the other hand, feel that one vineyard can be limiting and prefer to blend wines from various vineyards believing this brings complexity to the wine.

VINIFICATION
The processes used to make wine.

VINTAGE DATE
The year in which the grapes were harvested. For U.S. producers, 95% of the wine must have been harvested in the year stated on the bottle.

VISCOUS
In general it refers to the sensation of the weight of the wine in your mouth.

WINE

Also known as the drink of the gods, wine has been a part of the history of human kind for thousands of years. Someone once said that man merely existed prior to the discovery of wine and after that man learned to live. Wine is a wonderful expression of the earth and the very spirit of wine is one of enjoyment and pleasure. Have fun discovering wine and remember, it is your mouth, make it happy. A great wine is a wine enjoyed.

YEAST

Another variable in the process of winemaking. Yeast is a single-celled, airborne fungi of which there are many species and strains. Transported by insects, (like those obnoxious fruit flies that end up in your wine glass) or anything else that moves, these spores find their way into the vineyard and into the winery. Fermentation can not happen without yeast, the question is which yeast; natural or cultured. Natural yeast is a combination of many strains of yeasts that come in with the grapes from the vineyard. Cultured yeast is, typically, a single strain of yeast grown in a lab. Some yeasts affect the temperature and pace of fermentation, while others influence flavors and aromas. Some winemakers prefer to let the natural yeast do its work, while others choose to inoculate their juice with one or several types of cultured yeast, relying on the predictability and control that cultured yeast offers. This is a simplistic approach to something that is very important in the winemaking process. Further reading recommended.

Blending Varieties

Blending is a common practice and can incorporate many different techniques. For example, a wine can be a blend of various vineyard sites with each vineyard relating its own characteristic to the aromas and flavors of the wine. Some winemakers may age a certain percentage of the wine in new oak and another part in older, relatively neutral barrels in an effort to control the influences the oak imparts to the wine, later blending those wines before bottling. There are many types of blending, but we generally think of blending as the combination of different grape varieties. Whether it is a blend of grapes, vineyard sites, separate vinification processes, or finished wine from different barrels or steel tanks, all becomes part of the personality of the wine.

The varieties included here are the ones used most often to blend with Zinfandel. These varieties are selected for the character they will bring to the wine. Sometimes though, as one winemaker put it when asked why she blended a certain grape variety with her Zinfandel, "because it's there in the vineyard, growing along side the Zinfandel, a companion in the field, a partner in the wine."

Much has been written about each of the varieties introduced. I encourage you to continue researching them (and tasting them) to understand more about their individual characteristics, and their relationship to the regions and vineyards in which they are grown.

Alicante Bouschet (ah lay kan tay / boo shay)

Alicante Bouschet is a unique French hybrid with a crossbreeding of Grenache and Petite Bouschet. It is a teinturier grape variety, meaning the flesh as well as the skin is red. It is used chiefly as a blending grape but varietals are produced. Fairly adaptable to varying growing sites, it prefers a warmer climate and requires little if no irrigation. Capable of high yields, this red meat grape grows in moderately tight clusters with medium sized berries that have a thick, tough skin. Alicante Bouschet can be high in alcohol and have aggressive tannins, producing a full bodied, robust wine. Partnered with Zinfandel, this variety brings color, tannin structure and earthy notes to the blend.

Common characteristics: Smoky, earthiness, raspberry.

Further study: Garnacha Tintorera, Tenturier

Cabernet Sauvignon (cab air-nay so-vee n'-yohn)

Cabernet Sauvignon does well in most wine growing regions. A slow ripening grape, it performs best in warm temperatures and well-drained soil. As long as it can mature, cooler climates will produce grapes consistent with the basic characteristics of the variety. Cabernet Sauvignon is an extremely vigorous vine producing small, thick skinned, black berries with high tannin. When blended with Zinfandel, Cabernet Sauvignon adds to the tannin structure producing a wine that will continue to develop with age. Cabernet Sauvignon provides characteristics such as cherry, currant or cedar qualities that blend well with Zinfandels distinctive fruit flavors.

Common characteristics: Cedar, tea, raspberry, plums, chocolate, black cherry, currant, cassis, mint, violet, herbs, olives, green pepper or tar.

Further Study: Bordeaux

Carignane also spelled Carignan (care in yahn)

Budding and ripening late, this medium skinned, tightly clustered variety is not always easy to grow. It is a vigorous vine, sensitive to mildew and prone to rot, preferring a hot, dry site with good exposure to sunlight and well drained soil. Typically blended, Carignane provides high acidity, good tannin and deep color. As a varietal it is often dismissed, but some winemakers believe and have proven that handled properly it can produce wines of great depth and intensity. Old vines exist in California with one vineyard in Geyserville nurturing vines that were planted in the 1880's. Zinfandel can benefit from Carignane's color and tannins.

Common characteristics: Spice, dark, ripe fruitiness

Further study: Mazuela, Languedoc-Roussillon

Cinsault also spelled Cinsaut (san-so)

A prodigious producer, this medium sized, purple berry has a preference for hot climates. Typically high in acid, low in tannin and deep in color, Cinsault is an amiable grape for blending. Rarely seen as a true red varietal in California, it is commonly made into a dry rosé. Not difficult to grow and accepting of most soil conditions, Cinsault is often found in the field blend. It is highly aromatic with distinctive soft rose aromas.

Common characteristics: Almond, hazelnut, perfume, rose

Further study: Rhône, South Africa, Pinotage

Grenache (greh nah' sh)

A medium sized, reddish purple berry that tends to do best in hot, dry vineyards. It is the second most widely planted variety in the world. In the Rhône region, Grenache is capable of producing intense red wines that require cellaring, but grown in California it generally lacks tannin and color and is rarely produced as a varietal. Readily accessible and fairly easy to grow, Grenache is a most desirable blending grape. High in alcohol and sugar content, it brings a sweet berry fruitiness to the blend. Depending on the region Zinfandel is grown, blending Grenache can accent the characteristic berry flavors and soften the tannins.

Common characteristics: Strawberry aromas, sweet berry notes.

Further study: Cotes du Rhone, Châteauneuf-du-Pape, Rhone Valley, Navarra.

Merlot (mair-lo)

A thin skinned variety that buds and ripens early, Merlot was introduced into the vineyards of California in the early 1960's. The vines produce loose clustered, medium sized black berries that have a propensity for warm, dry climates but does well in areas as far north as Washington State. Often blended with Cabernet Sauvignon to soften tannins and round out the fruit, more robust Zinfandels may too benefit from Merlot's soft touch. Merlot is produced as a varietal and on its own, depending on where it is grown, can produce wines of depth and complexity. It has a medium to deep red color and a variety of fruity aromas lending towards cherry and currant.

Common characteristics: Herbal, cherry, currant, tea, orange rinds, and plums.

Further Study: St-Emilion, Bordeaux and Washington State.

Mourvèdre (moor-Veh-druh)

This small, sweet, thick skinned berry adepts well to a wide variety of soils but prefers a warmer climate as it tends to break bud and ripen late. Often high in alcohol and tannin, Mourvèdre, or Mataro as it is sometimes called in California, is typically used as a blending grape. Adapting well to most soils Mourvèdre can display a range of personality from earthiness to a full, black cherry and plum character. To the Zinfandel blend, Mourvèdre will convey these vineyard site qualities while adding color, tannin structure and a full mouth feel.

Common characteristics: Spicy, peppery, plum flavors, blackberry aromas.

Further study: Rhône Valley, Spain.

Nebbiolo (neh-b'yoh-lo)

Not widely grown in the U.S., this small berried grape can be rather fussy. Yields are typically low, and being a late ripener it requires a long, warm growing season. It has a preference for certain soil types and site locations, performing best on steep hillsides with fog intrusion necessary to develop all the aspects of this variety. (Nebbiolo is believed to be derived from the Italian word, nebbia, or fog.) High in acid and tannin, with color that is deep and dark and often part of the field blend, Nebbiolo is rarely produced as a varietal in California. But with the increased interest in Italian varieties, that may change. The rich texture and sometimes chewy qualities of Nebbiolo blend well with Zinfandels briar and spice character.

Common characteristics: Violets, roses, chocolate, licorice, dark berry fruit.

Further study: Piedmont, Italy

Petite Sirah (peh-teet sih-rah)

A black, medium sized berry with relatively thin skin, Petite Sirah ripens mid-season and can be prone to sunburn or raisining. Preferring a warmer climate it adapts well to most soils but does best in well drained and not overly fertile sites. The flavors, color and tannins of Petite Sirah are intense and the wines produced from this variety are often described as muscular. Long recognized as a valuable blending grape, Petite Sirah, with its high alcohol and chewy tannins, enhance the body and longevity of the wine. Distinct leather, spice and pepper distinguish this variety but notes of plum and cherry are often found, if not subtly. Petite Sirah produces a deep, inky colored wine bolstering and adding depth to Zinfandels hue. Zinfandel also benefits from Petite Sirah's brawny tannins, firming up the structure and bringing balance to the wine.

Common characteristics: Spice, pepper, tar, cedar, plum, cherry.

Further study: Durif, Southern Rhône.

Sangiovese (san-gee-o-vay-see)

Its birthplace is Italy where it is the dominate grape in Chianti. With its ancient history and roots in Italian culture it is of no surprise that immigrants would select this variety to plant in their new home in California. Sangiovese is difficult to define even in broad terms due to the amount of clonal selection associated with this variety. In the most general of terms, Sangiovese can be high in tannin and acid, and tends to be more earthy than fruity. It lacks any notable color but a tell-tale orange halo-like rim is often found in wine that is blended with Sangiovese. Preferring a warm climate, the relatively thin skins make it prone to rot. Not overly vigorous it adepts well to most soil types but performs best in well drained soils. It would seem that Sangiovese is more a part of the field blend when it comes to Zinfandel. As a varietal in California, and especially in Napa, it is slowly finding its way with the help of stubborn growers and tenacious winemakers alike.

Common characteristics: Barnyard, earthiness, leather, touch of spice and plum.

Further study: Italy, Chianti, Brunello and Montepulciano.

Syrah (see-rah)

Small to medium sized, and ovoid in shape, this black skinned berry is quite productive. A late budding variety, it requires a warm climate to reach full maturity and develop its rich, full flavors. It is the most prestigious grape of the northern Rhône region and has become one of the most widely planted grapes in Australia, where it is known as Shiraz. Appreciated for its longevity, Syrah produces full bodied red varietal wines and brings deep color, tannin, spice and structure to a blend.

Common characteristics: Spice, pepper, blackberries.

Further Study: Hermitage, Northern Rhône, the Rhône Rangers.

Other varieties

Various other varieties are found in the field blend, such as Rubired or Muscat, and you are encouraged to read about them. Typically they are considered a "lesser variety" contributing little if anything to the wine. Just a filler, they say. But somebody planted them in those old vineyards. Maybe they thought those varieties might add to the spectrum of color and flavors, even if it was so very subtle. There could be an aroma that your not quite sure what it is, some fleeting hint of sweetness or fleck of spice you just can't name, and it just might be those varieties, quietly capturing your attention.

Viticultural Areas

American Viticultural Areas (AVA)

The influence of soil, climate and topography on each growing region has long been recognized by grape growers and winemakers. In an attempt to clarify these regions and define their unique growing conditions for consumers, the BATF: Bureau of Alcohol, Tobacco and Firearms, established a system of identification called AVAs: American Viticultural Areas.

What determines a viticultural area is, in concept, its distinction from other growing regions. According to the rules of BATF, a viticultural area is, by their definition; "A delimited grape growing region distinguished by geographical features, the boundaries of which have been recognized and defined." These sets of regulations came about in 1978 with the purpose of setting labeling standards to aid the consumer with product information and to assist wineries in distinguishing their wine from others. It can be very confusing with any differentiation lost in the limitations of this system. But, for now, it is all we have.

AVA status is granted by permission of the BATF. Each growing region must petition the BATF and provide environmental and historical documentation to establish their distinction from other areas. Wine producers will declare the state, county, viticultural region or vineyard designation on their labels. For state of origin, all of the grapes must have come from that state. In California, listing the county requires that 75% of the grapes must have been grown in that county unless the county is designated as an AVA, then it becomes 85%. If a vineyard site is on the label then 95% of the grapes must come from that vineyard. (Unless it is designated as an "Estate" and then it must be 100%.)

One of the standards used to determine whether a growing region warrants AVA status is climatic diversity. In the 1930's a system of measurement was established by Maynard Amerine and A.J. Winkler, two professors at the University of California. They studied the typical growing season of vines, April through October, and determined that each growing region could be differentiated based on the measurement of temperature in a given day. Taking the highest and lowest temperature, adding them, dividing them by two then subtracting 50 equals degree days. The result is five separate regions.

REGION I: Fewer than 2,500 degree days. Climate typically suited to Johannisberg Riesling, Pinot Noir and other varieties.

REGION II: 2,500 to 3,000 degree days. Climate typically suited to Cabernet Sauvignon, Sauvignon Blanc, Merlot and other varieties.

REGION III: 3,000 to 3,500 degree days. Climate typically suited to Zinfandel, Petite Sirah, Syrah and other varieties.

REGIONS IV AND V: 4,000 degree days. Climate typically suited for red varieties, but can often be too hot to produce grapes of distinguished wine quality.

Coastal Cool and Coastal Warm

Another method used to determine climatic diversity for purposes of defining a growing region is the establishment of "Coastal Cool" and "Coastal Warm" by Robert Sisson of The University of California. This system of measurement was designed to supplement "degree days" and basically takes into account the number of hours at which temperatures remain in the photosynthesis range of 70°- 90°F range. Coastal Cool and Coastal Warm will reflect the amount of fog intrusion, with Coastal Cool having more fog influence than Coastal Warm.

The attempt at defining growing regions in relation to grape development is simplistic at best, for it does not address the internal diversification of an appellation. The reality is, that even within a single vineyard, variances can be identified. With that in mind, what is presented here is an overview of appellations and sub-appellations.

Viticultural Areas in California

ALEXANDER VALLEY

Named after Cyrus Alexander who settled in the area around 1840, Alexander Valley is located in the northern portion of Sonoma County, bordered by Mendocino County at its most northern tip with Dry Creek Valley its direct neighbor to the west. The Russian River flows through the valley with the summer fog from the Santa Rosa Plains captured in the narrowing of the southern portion of the valley. The soils tend to be gravely, sandy loam, well drained and often quite fertile. The soil warms early in the growing season and the vines are vigorous with yields high. Much of the vineyards are planted on the valley floor, river terraces and lower slopes rising out of the valley. Overall, Alexander Valley is considered to be a Region III, coastal warm area although other areas of the valley are categorized as Region I and II. Zinfandels tend to be described as "supple" with plump blackberry and black cherry flavors.

AMADOR COUNTY

Located in the Sierra Foothills, 40 miles east of Sacramento, Amador County has a history rich in gold and agriculture with some of the earliest documented Zinfandel vineyards planted sometime between 1852 and 1869. In the vineyards of Amador County, Zinfandel dominates. The area is susceptible to late spring frosts, has cold night temperatures that give in to very warm summer days and rainfall tends to be slightly higher in this area. Due to the high elevation of this area, fog is not a factor lessening the threat of mold or fungal disease. Considered a high Region III - IV, winds from the San Francisco Bay temper the area during the day with mountain breezes cooling the night.

Contra Costa County

Areas of Contra Costa are part of the Central Coast AVA, an enormous appellation that extends south of San Francisco to north of Los Angeles. Contra Costa County is located at the inlet of San Francisco Bay. There are some old vineyards in this delta region dating back to the turn of the century. The well drained sandy soils protect the vines from Phylloxera which can not live in this type of soil. The area of Contra Costa falls in the Region III to IV.

Dry Creek Valley

Dry Creek Valley, located in north central Sonoma County, follows Dry Creek, a tributary of the Russian River, from Healdsburg up towards Geyserville. Dry Creek Valley is sixteen miles long and 2 miles wide. The valley floor is fertile with well drained soils and moderate morning fog influence. The hillsides and benchlands are dominated by red soil with little daytime fog intrusion resulting in slightly warmer temperatures. It is late afternoon and evening that the moderating fog effect reaches the valley, cooling the grapes and allowing them to mature more slowly. Overall the Dry Creek Valley appellation is regarded as a warmer growing region, falling into the Region II - III and Coastal Warm categories. The hillsides and benchlands of Dry Creek Valley are ideal for Zinfandel having been planted here for over a hundred years with some of the old vines still existing and producing fruit. Zinfandel from Dry Creek Valley has a distinct core of raspberry and blackberry flavor and aroma.

El Dorado

With elevations ranging from 1,200 feet to 3,500 feet this AVA in the Sierra Nevada's Foothills has great diversity in geology, soil textures and content that accommodate a wide range of varieties. Similar to its neighbor to the south, Amador County, El Dorado lacks the cooling influence of fog intrusion but the mountain and bay breezes temper the area. A Region II to III, the warm days and cool nights are ideal for viticulture and rainfall occurs mainly in the winter months when the vines are dormant. El Dorado has a long viticulture presence with wineries established in the late 1800's.

HOWELL MOUNTAIN

A sub-appellation in Napa County, northeast of St. Helena, Howell Mountain has been the site of vineyards for over a hundred years. Elevation ranges from 1,400 feet to 2,200 feet where it is not fog that cools the vineyards, but wind from the Pacific. The soil is volcanic in origin with warm red soils dominating the north face of the mountain and lighter gray ash based soils on the south side. It is a Region III, Coastal Warm appellation.

LIVERMORE VALLEY

The Livermore Valley, first planted in the 1880's, lies 50 miles east of San Francisco, encompassing an area fifteen miles long and ten miles wide. Mountains surround the valley and early morning fog from the San Francisco Bay influence the area throughout the growing season. In the late afternoon, cool marine breezes push the coastal fog back into the valley. The soil is rocky and well drained with a climate considered to be a warm, low Region III.

LODI

Located in the immense Central Valley AVA, Lodi is situated between the cities of Stockton and Sacramento. Parts of the Central Valley can reach a Region IV, but the sea breezes moderate certain areas of Lodi, making it a Region III. The soils are sandy and well drained in most of the region. Lodi stretches into the Sierra Nevada and the rocky, shallow terrain of these foothills provide a good site for Zinfandel.

MENDOCINO

Mendocino County includes Anderson Valley, McDowell Valley, Potter Valley, Cole Ranch and Redwood Valley AVA's. An extremely diverse growing region that moves inland from the ocean, through redwoods into canyons and valleys. Generally a Region III, areas of Mendocino, such as Anderson Valley, which is a Region I to II, are greatly impacted by fog but other sections see little if no fog and are tempered by the mountains.

MOUNT VEEDER

Located on the eastern slope of the of the Mayacamas Mountain range, Mt. Veeder is just west of the city of Yountville. Rising 2,500 feet above the Napa Valley floor, temperatures are warmer at night and cooler in the day then the vineyards located in the valley. Like most mountain vineyards, the amount of sun exposure depends on location with sunlight slipping into the shadows of gullies leaving other sites still basking in the suns rays after dusk has crept into the valley. Soils take on a broad scope with a variety of soil types and textures. Rainfall is heavier than the valley but fog is rarely a factor with the vineyards cooled by the mountain breezes.

NAPA

Extending from the San Francisco Bay, where the Napa River ends its journey, northward to the base of Mt. Saint Helena, Napa is considered one of the most diverse growing regions in California. With its broad range of topography, soil and climatic influenced, a multitude of grape varieties thrive and flourish there. To the south, the cooling conditions of the bay embrace heat-sensitive varieties such as Pinot Noir where in the mid and northern part of the valley, the almost tropical-like temperatures are renown for inspiring Cabernet Sauvignon. Zinfandel seems to enjoy the higher elevations of Mt. Veeder and Howell Mountain where it can revel in the sunlight. Napa includes the sub-AVA's of Atlas Peak, Howell Mountain, Los Carneros, Mount Veeder, Napa Valley, Oakville, Rutherford, Spring Mountain District and Wild Horse Valley.

NORTHERN SONOMA

This AVA embraces Alexander Valley, Dry Creek Valley, Knight's Valley, Russian River Valley, Chalk Hill and Green Valley. This AVA does not often appear on labels, but if it is used then it means the grapes could have come from two or more of these regions.

Oakville

Located between Rutherford and Yountville, Oakville is considered a Region III with the cooling effect of fog occurring in the early mornings. Part of the greater Napa AVA, Oakville has a long viticultural history dating back to the 1870's. Cabernet Sauvignon is king here but Zinfandel is no pauper, producing rich, well balanced wines.

Paso Robles

"El Paso de Robles" or the Pass of the Oaks, as travelers from the mission of San Miguel to San Luis Obispo once called it. Located in the county of San Luis Obispo, Paso Robles has a long viticultural history. Despite its proximity to the ocean (30 miles or so), the Santa Lucia Mountain range blocks any benefits the Pacific has to offer. Cooling breezes, as slight as they may be, come from the Monterey Bay via the Salinas Valley. Rolling hills, 750 feet to 1,800 feet elevation, characterize the area. A Region III, Coastal Warm appellation, with cool nights and warm days, Paso Robles is ideal for the sun loving Zinfandel as well as Cabernet Sauvignon.

Sierra Foothills

What was once gold in "them thar hills" is now Zinfandel, by far the most widely planted variety in this area. Amador, Calaveras, El Dorado, Mariposa, Nevada, Placer, Tuolumne and Yuba are all part of the Sierra Foothills AVA. Some of the oldest vineyards in the state exist here, protected by their relatively remote location. Although Zinfandel dominates, other varieties are successfully grown here.

Sonoma Coast

Part of the greater Sonoma AVA, Sonoma Coast includes sections of Sonoma Valley, Green Valley, Carneros and Russian River Valley. The primary distinction of this AVA is that the areas included are limited to Region I to II, Coastal Cool growing regions.

SONOMA COUNTY

An expansive AVA that incorporates Alexander Valley, Chalk Hill, Dry Creek Valley, Knights Valley, Los Carneros, Northern Sonoma, Russian River Valley, Sonoma Coast, Sonoma County Green Valley, Sonoma Mountain and Sonoma Valley AVA's.

SONOMA MOUNTAIN

A sub-appellation of Sonoma Valley, the vineyards of Sonoma Mountain rise above the fog line experiencing warmer temperatures then the valley floor. Temperatures remain fairly stable throughout the growing season and rainfall tends to be slightly higher. Zinfandel and Cabernet Sauvignon enjoy this sunny perch but the Region I to Region II climate is mild enough for heat sensitive varieties as well.

SONOMA VALLEY

Stretching from the San Pablo Bay and a few miles beyond the city of Kenwood is the Sonoma Valley AVA. Mountains, foothills, plains and valleys all lay within the boundaries. With such diversity, this area is home to wide range of varieties. San Pablo Bay and the Petaluma Gap, a break in the coastal mountain ranges, bring in the fog and cooling breezes that moderate the temperatures. Summers are somewhat cooler, winters warmer with rainfall lower than in most regions. Sonoma Mountain lies to the west and Mayacamas Mountain range to the east with the town of Sonoma secure in between. Soils run the gambit with volcanic soils dominating. Sonoma Valley is considered a broad Region I to III with temperatures warming the further north you go. This area has enjoyed a long viticultural history having been planted in the early 1820's. Zinfandel thrives here producing fruit in a broad range of berry flavors, from blueberry to cranberry to dark raspberry with often a touch of mint or chocolate.

Redwood Valley

Recently granted the status of AVA, Redwood Valley is in the greater appellation of Mendocino. Redwood Valley begins six miles north of Ukiah, extending ten to twelve miles south, following the course of the Russian River. Rolling hills and rugged terrain typify this region. The area is slightly cooler than Ukiah Valley with a vast array of microclimates.

Russian River Valley

A growing region which is tempered by the cooling effect of the Pacific Ocean, the Russian River and afternoon winds. Considered a Region I, Coastal Cool appellation, the Russian River Valley stretches from the town of Healdsburg extending south and west to the city of Guerneville. Zinfandel is planted primarily in the northern area of the appellation where temperatures are slightly warmer. During the growing season the fog intrusion can be very substantial and tends to remain through the early morning when the sun manages to gain enough energy to burn it off. The growing season in the Russian River Valley is a long one providing a slow, steady ripening of the grapes. This extended hang time intensifies the flavors in the fruit producing very lush, blackberry, raspberry characteristics in the grapes.

Index

Sullivan Ranch, 23
Sunset Cellars, 291–93
Sutter Home Winery, 1, 181–82
Swan, Joseph, 1, 135
Swan, Joseph Vineyards, 135–37
Swiss Collina Vineyard, 64

Taft Street Winery, 294
Talbott vineyard, 183
Talmage Bench, 93
Tambollini Vineyard, 209
Tarius Wines, 295–97
Taylor Ranch, 25
Tchelistcheff, Andre, 21, 162
Tchelistcheff, Dimitri, 162
Tcholakov, Miro, 302
Teldeschi, Dan, 88
Teldeschi, Frank, 88
Teldeschi, F. Winery, 88–90
Teldeschi, Ray, 36
Teldeschi Vineyard, 88, 89, 221
Teller, Otto, 219
Templeton, California, 193
Templeton Gap, 193, 195
Templeton Hills, 155
Ten Point Vineyard, 141–42
Terra d'Oro Zinfandels (Montevina), 180–82
Tierney, John, 294
Tobias Vineyards, 193
Tobin James Cellars, 300–301
Trentadue, Evelyn, 234, 302
Trentadue, Leo, 234, 302
Trentadue Ranch, 233, 302
Trentadue Winery, 302–3
Tria, 304–5
Trinchero family, 178, 182
Trotta, Linda, 120
Truchard, JoAnn, 306
Truchard, Tony, 306
Truchard Vineyards, 306–7
Tuolumne County, 346
Twin Rivers Vineyard, 55, 56, 285

Ukiah, California, 93, 296, 297
Ukiah Valley, 78, 347
Ullom, Randy, 143
Underwood family, 84
Unti, George, 308

Unti, Kimberly, 308
Unti, Linda, 308
Unti, Mick, 308
Unti Vineyards, 308–10
Upper Weise Ranch, 146–47
Upton, Lloyd, 140

Van Staaveren, Don, 270
Vesuvium Zinfandel (Schuetz Oles), 273–74
vineyard designations, BATF labelling rules, 341
vintage of 1998, 7–8
viticultural areas, 341–47
viticultural terminology and techniques. See Glossary
Vix family, 84
Vukic Vineyard, 268

Walker, Lloyd, 32
Walker Ranch Vineyard, 31–32
Wann, Grady, 208
Warren, Kevin, 23
Warrior Fires Zinfandel (Karly), 139
Wasserman, Fred, 42
Wasserman, Pam, 42
Watts Vineyard, 20
Wendt, Henry, 208
Wendt, Holly, 208
Westberg, Tom, 193
Westside Paso Robles, 63, 154, 155, 195
Whaler Vineyard, 93–94
White Cottage, 267
White Oak Vineyards and Winery, 311
white Zinfandel, 3–4
Wild Hog Vineyard, 314–15
Wild Horse Valley, 345
Wildwood Ranch, 149
Williams, John, 104
Williamson, Van, 77
Wine Creek Vineyard, 111
winemaking terminology and techniques. See Glossary
Winkler, A. J., 341
Wooden Valley Vineyard, 216
Wood Road Vineyard, 91, 92